BASIC ROUTINES FOR MASSIVE MUSCLES

BASIC ROUTINES FOR MASSIVE MUSCLES

ROBERT KENNEDY

Sterling Publishing Co. Inc., New York

**Library of Congress Cataloging-in-
Publication Data Available**

10 9 8 7 6 5 4 3 2 1

Published by Sterling Publishing Company,
Inc.
387 Park Avenue South, New York, N.Y. 10016
Revised edition of *Beef It!* published in 1983
by Sterling Publishing Company, Inc.
© 1983, 1998 by Robert Kennedy
Distributed in Canada by Sterling Publishing
c/o Canadian Manda Group, One Atlantic
Avenue, Suite 105
Toronto, Ontario, Canada M6K 3E7
Distributed in Great Britain and Europe by
Cassell PLC
Wellington House, 125 Strand, London WC2R
0BB, England
Distributed in Australia by Capricorn Link
(Australia) Pty Ltd.
P.O. Box 6651, Baulkham Hills, Business
Centre, NSW 2153, Australia
Printed in Hong Kong
All rights reserved

Sterling ISBN 0-8069-7761-2

CONTENTS

Foreword 6

1 **Motivation and Setting Goals 8**
Defining Your Desires

2 **Somatotyping 14**
Do You Need It?

3 **The Training Log 20**
The Silent Reminder

4 **Machines or Weights? 24**
Which Are Better?

5 **Reps Under Scrutiny 30**
Challenging Muscle-Fiber Contraction

6 **Injuries 36**
How to Avoid Them

7 **Cycle Training 40**
Pushing to a Peak

8 **Power Thinking 44**
Mental Programming for Success

9 **Super-Structuring Your Routine 48**
Improving Your Level of Efficiency

10 **Recuperation 54**
Mending the Muscle

11 **Body-Fat Percentage 62**
The Lean Advantage

12 **Metabolism Training 66**
Creating the Anabolic State

13 **Ultimate Nutrition 70**
Muscle Building and the Food Factor

14 **Derailing the Sticking Point 78**
Regulating the Physiological Processes

15 **The Muscle Sleep 82**
Snoozing for Size

16 **Shoulders 86**
Building Barn-Door Width

17 **Chest 94**
Sculpting the Pecs

18 **Abdominal Training 102**
Wasting Away for Midsection Impressiveness

19 **Putting Your Back into It 108**
Working the Lats and Traps

20 **Quad Training 116**
Maxing Out the Upper Legs

21 **Calves 122**
Building the Lower Legs

22 **Arms 126**
Filling Out Your Sleeves—but Quick

23 **Forearms 136**
Bringing the Lower Arms Up

24 **Tanning Up 140**
Natural and Artificial Methods

25 **State-of-the-Art Supplements 146**
An Overview of the Latest

26 **Routines for Adding Size 158**
Maxing Out the Muscle

27 **Tips to Beat Out the Competition 168**
The Winning Edge

28 **Posing 174**
The Art of Successful Physique Display

29 **The 10 Bodybuilding Pitfalls 180**
Small Leaks Can Sink a Battleship

30 **Questions and Answers 186**
Help!

Index 190

FOREWORD

Scott Wilson

Bodybuilding seems truly magical. A few months devoted to weight training can alter your body's composition, shape, fitness, and strength to an incredible degree. Combine your workouts with correct nutritional intake, and the results can be really amazing.

None of the athletes who appear in the pages of this book had huge muscles before they started working out. Some were downright weak and skinny. Many were overweight; still others started life with less than robust health. But for one reason or another, each discovered the sport of bodybuilding—and took it to the max.

Whether or not you aspire to be a competitive bodybuilder, you will have to train following the same basic principles that these professionals use in their own training. This translates to working out with weights three to five days a week, each session lasting from a half hour to an hour. You will agree, I'm sure, that this is no big sacrifice. The other essential aspect has to do with nutrition. To gain muscle, you will have to eat more. To lose fat, you will have to eat less. For those who want to gain muscle while losing fat . . . well, the answer lies within the pages of this book. It involves a little bit of nutritional balancing, but it's not rocket science, and it's certainly not difficult to incorporate into your lifestyle.

My own introduction to bodybuilding came at the tender age of 10. My mother had taken me to the Drury Lane Theatre in London to see *South Pacific*. I enjoyed all the songs, but what really captured my attention was the astounding physical development of one of the sailors in the cast. I just couldn't believe my eyes. The guy was built from head to toe. Wow! Did I ever admire that guy's physique. How on earth did he get like that? Could I ever hope to be even slightly muscular myself someday?

It turned out that the *South Pacific* sailor was none other than London bodybuilder Stan Howlett. Today, he is 70 years of age, fit and trim, and—amazingly—still competing in bodybuilding contests. After seeing Stan in *South Pacific*, I waited to meet him outside the stage door of the theater. From quizzing him nonstop, I learned that he had built his muscles with "progressive resistance exercise." I thanked Mr. Howlett, but in truth I was still in the dark about how I could build up my own skinny body.

Martin Marek

It wasn't until a full year later that I learned that progressive resistance exercise was another name for weight training. Now I not only knew what I wanted, but I knew the path to get there. But I had one problem: no money to buy weights. I just could not afford them. Still, my motivation was so hot that I got the idea to search the local dump, railroad tracks, and wrecked-car yards for something I could use instead. I found two large metal biscuit containers and a long steel bar. I filled each of the containers with rocks and, after inserting the bar, poured cement over the top. Presto! I now owned my very own barbell. Over the next few months, I created three more of these monstrosities. But however crude these barbells looked, believe me, they were my most prized possessions. They were my lifeblood! I was now a practicing bodybuilder.

This book is for everyone who wants to improve his or her physical appearance. Weight training is unequivocally the fastest way to build up your body. Other benefits are more robust health, increased strength, and improved fitness. Not a bad trade-off for a few short workouts each week.

Remember that you can take bodybuilding as far as you want. This book is for the underweight individual who wants to firm up and fill out all the way to the competitive bodybuilder who is interested in pumping iron, ripping up, and beefing it. All adhere to the same basic training principles. So, how are the different levels achieved? It comes down to training intensity, nutrition, and supplementation—subjects thoroughly discussed in this book.

This revised edition of *Beef It!*, published in 1983, contains an all-new chapter on the latest developments in supplements, from antioxidants to natural testosterone boosters. If you want to know what bodybuilders the world over are talking about at the gym these days, be sure to check out State-of-the-Art Supplements (Chapter 25).

Also within the pages of this book, you will learn about super-structuring your routine, controlling your metabolism and ridding your body of excess fat, derailing sticking points, setting goals, and contest preparation—all interspersed with tips from the champions.

If you train diligently and follow the suggestions in this book, you will be able to work wonders!

1

MOTIVATION AND SETTING GOALS

DEFINING YOUR DESIRES

M otivation is what drives us forward, more than anything else. If you have it— whoopee!—nothing can hold you back. The gas pedal is locked in the down position, and all you have to do is steer.

I well remember my younger days when body-building was the major focus in my life. Nothing else mattered. Reading? No way—unless it was about building muscles. Conversation? Only on the subject of pumping iron and working out. During those glorious misspent days, I would turn down dates, wedding invitations, trips, hikes, anything, all because of the possibility of missing an almost sacred workout. How I cursed the bodybuilding magazines back then for insisting on training only three times a week. Why, I could have trained six hours seven days a week!

If you are as driven as I used to be, you could become a victim of your own enthusiasm. You could overtrain, become totally disillusioned, and ultimately give up in disgust. But it's far better to be overinspired and overmotivated than not to be inspired or motivated at all.

When that fire and longing burn deep within your bones, it is never a matter of whether you should train or not. The question simply doesn't arise. You wouldn't miss a workout even if you were paid to. The only question is how to get the most out of your workouts.

Some believe that the successful bodybuilders, the top champions, are enthusiastic and obsessive to the extent of being emotionally disturbed. Food for thought, indeed. On the other hand, publisher Joe Weider, in *Joe Weider's Flex* magazine, states: "I have always believed that people who take up bodybuilding are the doers of the world. Given an outlet for their energy and creativity, they will build monuments." Whatever the case, thank God for your motivation. Channel your drive carefully, however. You must temper it with intelligent application to correct training. Otherwise, Burnout City!

Still, it's wise to take advantage of the drive you have now, because chances are that as you mature (heaven forbid!) your drive will gradually diminish. When motivation goes, you are in the unenviable position of having to drum up your enthusiasm artificially. You also might lose your attention span for training. Sure, you may still want a great-looking body, but wanting is very different from needing.

(Opposite) **Porter Cottrell**
(Right) **Frank Zane**

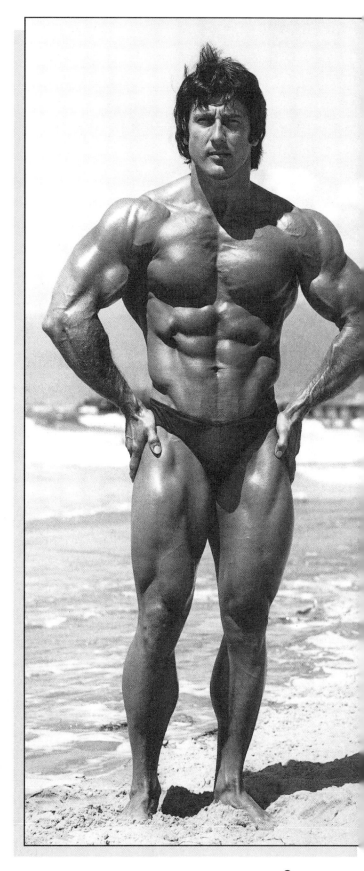

Sometimes a bodybuilder loses this aggressive desire when he finds a mate or achieves some kind of financial independence. Can you imagine saying, "Oh, no! I've been left a million dollars, and now I won't have the drive to make it as a top bodybuilder." Remember how tennis superstar Bjorn Borg lost his "killer instinct" when he got married? It doesn't always work that way, of course. Jimmy Connors got more fired up after his marriage.

It is almost inevitable, however, that your need for huge muscles will gradually diminish as the years advance. Committed bodybuilders will never lose this need completely, but, believe me, it lessens—enough to result in less assertive workouts, missed training sessions, and less discipline at the gym.

Regardless of whether you have an obsessive need or only a wholesome desire, without positive and clearly visualized goals you simply cannot succeed. If you are not already motivated, you must motivate yourself. Many people, bodybuilders included, remain satisfied with too little. Do not be content with insignificant achieve-ments. Many get just so far. Do not be content with developing a body that is just a little better than the average physique. Lack of drive is death. Apathy is your enemy!

As a bodybuilder, you cannot rest on your laurels. You must make every effort to cultivate a positive mental outlook. You must develop an attitude to life that does not accept permanent limitations. You need to possess a spirit of abounding enthusiasm and the determination that ultimately makes failure impossible. Whatever your goals, conceive them, believe them, and they will manifest. Forget the scoffers who say you are wasting your time. They are only revealing their own lack of spirit and may well be envious of yours. Forget those who say, "Even if you succeed, it's not worth it!" You have enough to handle without letting these negative thinkers weaken your resolve the least bit.

It can be helpful at times to feel a certain degree of dissatisfaction—not the dissatisfaction of despair, but a dissatisfaction with your present level of achievement. You can use this as fuel for forging ahead, trying harder,

Boyer Coe, Arnold Schwarzenegger, and Roger Walker

Lee Haney

and achieving success. Never fall into the trap of being satisfied with too little.

Bernarr MacFadden, known as the "Father of Physical Culture," was frequently heard to say, "Without a goal, you will never succeed." He should know. He was a physical marvel until well into his eighties, and by that time he had built up a publishing empire worth millions of dollars.

I like what *Iron Man* magazine publisher John Balik says about goals and attitudes: "Without clearly defined goals, short and long term, your journey through bodybuilding's waters will be like a ship without a destination. You must take the time . . . to clearly define your direction. . . . You have to understand your potential, and remember that every action starts as a thought. Without that first step, nothing is possible." According to Balik, you should ask yourself these three key questions:

1. What do I want out of bodybuilding? More weight? Leaner physique? Better proportions? Ultrafitness? To win Mr. Olympia?
2. What body type am I? An endomorph (heavyset and fat), a mesomorph (big-boned and muscular), or an ectomorph (small-boned and skinny)?
3. What is the level of my bodybuilding experience?

As a beginner, you should aim to bench-press your body weight six times, and squat with your body weight for 15 reps. As an intermediate, you should aim for one and a half times your body weight for 6 reps in the bench and for 15 reps in the full squat. When you get into advanced bodybuilding, you have to set your goals one month at a time. Write them down, and post them on your notice board or fridge, where you can see them every day. That will remind you continually of where you want to be. Reinforcing your goals on a regular basis will help you determine what actions you should take.

If your present training has not worked sufficiently, then instant change is what is needed. Remotivate yourself. Set realistic goals, and power your way with determination all the way to the top!

Arnold Schwarzenegger was one of the first bodybuilders to talk about visualization and setting goals. Speaking to Ken Dychtwald of *New Age Magazine*, Arnold gives us a glimpse into his unique way of thinking on the subject: "When I was very young, I visualized myself as being there (at the top of bodybuilding), having achieved the goal already. Mentally, I never had any doubts in my mind that I would make it. I always saw myself as a kind of finished product out there, and it was just a matter of following through physically.

"But mentally I was there already. It makes it so easy, because then when you train for hours a day, you don't question yourself anymore. 'What am I doing here?' You focus right in again on your vision, on your image of what you know you will be, and that's why you're in the gym each day, to get a step closer to your goal.

"When I trained my biceps, I pictured huge mountains, much bigger than biceps can ever be. Just these enormous things. You do something to your mind in order to do certain things. I know my biceps aren't mountains—although they may look like miniature mountains. But thinking that they are gets my body to respond.

"When weightlifters are standing in front of a barbell, they must, in their minds, lift it in order to then lift the weight physically. If they have even one percent of doubt in their minds, they won't do it. When they stand before the bar and they close their eyes, what they are doing is lifting the weight mentally. And if they fail to lift it mentally, then they won't make the lift at all. Sure they may try. They go through the motions so that they can't be accused of not having a go . . . but it comes down to the same thing. Grunts and groans and ultimately a missed lift.

"Before a workout I would flex my muscles to get in touch. I locked my mind into my muscle during training, as if I'd transplanted my mind into the tissue itself."

Closely related to the process of visualization is the art of positive thinking. There always will be people who will try to put you down or sabotage your efforts, either consciously or unconsciously. When people are on a diet, undoubtedly some kind friends will try to get them to eat. It can be the same with training. Someone is sure to come up with all sorts of reasons why you shouldn't go to the gym on a particular day. Friends like this you don't need. Surround yourself with positive, supportive people. Right now you are bound for success. See the trail ahead filled with golden light, and allow nothing to get in your way. Don't let a negative thinker derail your bodybuilding journey. Keep your head straight, and deviate for no one. Define your desires—then make an intention, and go for it!

Vince Taylor

Flex Wheeler

Paul Dillett

2
SOMATOTYPING

DO YOU NEED IT?

All men are created equal, or so it's been said. Certainly, many people assume this to be true, if for no other reason than they have heard it so often. But think of your various friends. Can you honestly say that they were all created equal? One may have a keener mind, another may have more natural strength, and yet another may possess better mental and physical attributes than the other two. Each of us comes into the world with a unique set of genes that endows us with some advantages and some disadvantages.

Way back some 400 years before the birth of Christ, Hypocrates broadly classified human beings into four different types: the choleric, the sanguine, the phlegmatic, and the melancholic. This classification was expanded in 1797 by the Frenchman Halle, but today most people use the physique classifications, or somatotypes, established by Dr. William H. Sheldon of Harvard University, a leading exponent of anthropometry.

After extended work with thousands of subjects, Sheldon concluded that there are only three distinct body types, although he acknowledged that most people are actually a combination of these types. The three physical types identified by Sheldon are endomorphs, mesomorphs, and ectomorphs, described below.

Endomorphs

Endomorphs have a rounded, pear-shaped physique and a tendency to be fat. They have a wide thorax and a long, rounded abdomen. Their small intestine is 23 to 25 feet long, and their large intestine 5 to 8 feet long. Thus, their food has a much greater distance to travel than it has with the other body types, so endomorphs derive more nourishment from their food. Most endomorphs have a placid, cheerful disposition, large bones, and a weight problem that probably started in childhood.

Mesomorphs

As far as bodybuilding goes, mesomorphs are the lucky ones. They are adept at strength sports and naturally muscular, and they have a strong, forceful appearance. They are staid, reliable, and normal in their habits and dietary requirements. In bodybuilding circles, they are known as "quick gainers."

Ectomorphs

Ectomorphs have egg-shaped heads and angular features, and most of them are bundles of energy. Long

(Left) Mike Mentzer
(Above right) John Grimek

and gangly, they usually don't have the slightest trace of fat. (Their small intestine has a length of only 10 to 15 feet.) Thin bones are their trademark, their rib cage is long, and at the thorax the angle is narrow. Having efficient lungs, they make natural long-distance runners.

Physiques of Champions

By now you probably are wondering whether you have to be a mesomorph to get to the top in bodybuilding. Actually, no Mr. Olympia, from the past up to the present, has been a pure mesomorph, but they all have a strong mesomorphic component in their makeup.

In real life, pure body types are rare. Most people are a combination of the three body types, although one of the types may be dominant. Therefore, Sheldon devised a numbering system for showing the degree to which an individual represents each of the three types.

The degree is expressed in numerals ranging from 1 to 7, with 1 denoting the lowest degree and 7 the highest. A person's somatotype is then expressed in three numbers. The first number denotes the degree of endomorphy (roundness and fat), the second number the degree of mesomorphy (muscle, bone, and strength), and the third number the degree of ectomorphy (fragile bones, high energy, and low body fat).

To illustrate this numbering system, let us consider the physiques of several famous champions.

Frank Zane

His mesomorphic rating is about 6 or 7. He is very low in ectomorphy. There is very little of the long-distance-runner type about him. He would rate no higher than 1 in ectomorphy. Grimek is probably a 3-6-1. He has a markedly dominant mesomorphic component, which could be one reason why he was never beaten in a physique contest.

Frank Zane
(Three-Times Mr. Olympia)

Zane is not in the least dumpy or pear-shaped, so his endomorphic component is obviously minimal. He gets a 1 in the endomorphic category. His muscularity has always been evident, even from the earliest days of his training. I rate him 6 in the mesomorphic category. As for his ectomorphic component, Zane shows some evidence of a fine bone structure, although it is not fragile. Let's rate him a 2. Altogether then, Zane approximates the somatotype 1-6-2. Again, the mesomorphy ranks very high.

Mike Mentzer
(Mr. Universe)

I give Mike a rating of 2 for his endomorphic component. He is far from having a pear shape, but his bones are very thick-set and he has a slight inclination toward "thick skin." His abdomen is long from the thorax. His mesomorphic rating, to my mind, is monumental. He must be a 7. Mike has muscles from head to toe. I think you will also agree that he shows very little evidence of the thin, nervous ectomorphic type. I rate Mike Mentzer a 2-7-1 somatotype.

Arnold Schwarzenegger
(Six-Times Mr. Olympia)

Rating Arnold is no easy task. He has the round head and general thickness of an endomorph, the muscle insertions and masculine excellence of a mesomorph, and some of the sensitivity of an ectomorph. I may be wrong, but I see Arnold as a 2-6-2.

Dorian Yates
(Multi IFBB Mr. Olympia Title Holder)

Dorian is a phenomenon of our times. The Britisher started out in bodybuilding as quite a thin individual, which is a strong indication of ectomorphic tendencies. He shows a shadow of endomorphy, but as is the case with most successful bodybuilders, the mesomorphic component is by far the most dominant. I rate Dorian as a 2-6-1.

John Grimek
(Two-Times Mr. America)

John shows a moderate degree of the first (endomorphic) component; he's probably a 3. He is undoubtedly very high in mesomorphy, having large, square bones (his wrist is well over 8 inches), excellent muscular attachments, and a high muscle-cell count. Even without training, he would have fairly large muscles.

16

Paul Dillett
(Top IFBB Pro Bodybuilder)

Paul has probably the most outstanding potential of any bodybuilder alive today. His seemingly natural propensity for muscle mass is awesome to behold. My rating for Paul's mesomorphic reading has to be the maximum. Incredibly, he has a sensitivity to his skeletal structure that gives him more than a hint of ectomorphy. I rate Paul Dillett as a 1-7-2. He has an amazingly strong mesomorphic component.

Hamdullah Aykutlu
(IFBB World Champion)

Hamdullah, who hails from Turkey, has a fine body, although it can never rival the massive size of the physique of a Dorian Yates or a Nasser El Sonbaty. Hamdullah is slimmer by nature than most competitive bodybuilders. His outstanding features are his beautiful shape and extreme cuts (definition). Even his glute muscles are striated from top to bottom. I rate him as a 1-5-3.

(Above) Hamdullah Aykutlu
(Left) Dorian Yates

17

Nasser El Sonbaty
(IFBB Mr. Universe)

Nasser is huge. When viewed from the front, he makes multi–Olympia winner Dorian Yates seem almost like an average competitor. Admittedly, the picture changes when the two men compare their backs. Nasser, who is originally from Egypt, now lives in Venice, California. He has gained more than 100 pounds of pure muscle since beginning weight training. I rate Nasser El Sonbaty as a 2-6-1.

The Uses of Somatotyping

You have seen now that it is not difficult to determine the somatotype of an individual. However, we cannot get an accurate picture of someone by just knowing the three numerals that denote his physical makeup. In other words, somatotyping gives us no more than a rough idea of a person's physical structure.

No doubt the question of whether you can change your somatotype through training or diet has come to your mind. I suspect that the official medical pronouncement would be that your basic type cannot be changed. However, in all the years I have been associated with bodybuilders and bodybuilding, I have had occasion to observe many structural changes that came about through regular training. There is no doubt that bones grow to accommodate muscle gains. I have also seen the basic shape of the thorax change after a few steady months of weight training.

Have you ever seen people who have lost a lot of weight? Even though you may never have seen them when they were overweight, you can still tell that they were once heavy. The reason for this is that their bones have not yet shrunk. Somehow, their bones look too big for the amount of weight they are now carrying. Bone growth during the period when you are gaining weight—whether from fat, muscle, or both—is relatively slow. Should you lose that weight, your bones as well as your skeletal muscles will be slow in returning to their previous smaller dimensions. This is all further evidence that structural changes actually do take place.

Many years ago, the bodybuilding fraternity was persuaded by an unending series of articles and advertisements in popular bodybuilding magazines purporting that an aspiring bodybuilder could not possibly make gains in muscular size and strength unless he knew his exact body type and tailored his training ex-

Nasser El Sonbaty

actly to his particular classification. Certain advertisements would require the bodybuilding student to submit (along with a sizable check) front, back, and side photographs, together with a photo depicting his abdominal retraction (his thorax with the waist sucked in to show the shape of the lower ribs). The more "straight across" the ribs appeared, the more endomorphic the pupil was; the narrower the angle, the more ectomorphic. Guided by that, the mail-order trainer would custom-design a schedule to get the maximum results for the specific body type.

It is my opinion that too much emphasis was placed on the importance of this body-type classification. Physique classification has several practical uses. It indicates an individual's hereditary makeup and therefore his potential for physical improvement. It also indicates the sport at which he is most likely to excel. However, it is not a key to exactly how a person should train (how many sets, reps, or exercises). Nevertheless, there are certain guidelines to which the specific types should adhere.

The Endomorph's Training

The endomorph needs to be motivated and kept enthusiastic, so he should surround himself with others when he is training. If he opts to train by himself at home, chances are he will end up watching television and eating a big bag of potato chips instead of hitting the weights.

Many endomorphs have a sluggish metabolism. Therefore, every session should contain some form of very stimulating exercise to step up his metabolism: high reps, squats, running, jumping rope, and the like.

Extreme endomorphs may, in fact, be suffering from a hormone imbalance. All of us have hormones and some of the characteristics of the opposite sex, but a man who has wide hips, round buttocks, a rounded lower abdomen, little body hair, and noticeable fat on the pectorals might want to consult his doctor about the advisability of getting some hormone therapy.

Needless to say, the endomorph, with his overefficient digestive system, must limit his overall caloric intake. He is, unfortunately, destined to be forever hungry. The food he eats should not be overcooked but as near to its natural state as possible. He also should avoid like the plague junk foods of all types, sugars, animal fats, and excess salt.

The Ectomorph's Training

Ectomorphs are slow gainers, and often their ambition exceeds their capacity. Once they embark on a schedule, they keep adding exercises to it until their routine takes hours to complete. They are constantly changing their workout, some trying a new schedule every week or two.

Ectomorphs use up oceans of energy, never sit still, gobble their food, go to bed late, and are frequently tense and worried. No wonder they find it difficult to gain weight.

The primary aim of this thin man impatient to become a muscle man is to slow down his metabolism by getting as much rest and relaxation as possible. Meals should be taken frequently; he should have five or six small meals a day. Smoking and alcohol are definitely out, for they rev up his already revved-up nervous system.

More than anything, the ectomorph must tailor his routine to his wiry musculature. He will not gain on endless sets and reps. The routine must be severe enough to stimulate muscular growth, but brief enough to prevent nervous drain.

At the outset, eight exercises will be sufficient—3 sets of 8–10 repetitions. He may add an additional set and a couple of exercises as time goes on. Intensity, too, can be increased, but only in relation to his experience and condition.

Warning: Daily training is definitely not recommended. Too much training is worse than too little. For the ectomorph, the key is knowing when to stop.

The Mesomorph's Training

I once knew a man who did nothing else but curl two 25-pound vinyl discs on a bar (total weight, 70 pounds) once a week. He did no triceps work, no other exercise, nothing. His arms were just over 19 inches. Another fellow, a truck driver, did no weight training, had never done any at all. He was the laziest man I ever met. His arms were 17 inches. Yes, both were mesomorphs—bodybuilding's "Chosen People."

Of course, even the purest mesomorph will not make it to the top in bodybuilding just by his superior genetics. That may have been possible years ago, but today it takes more than mere heredity. There are so many variables: posing, diet, proportion, charisma—the list is endless. Today, even the mesomorph has to train hard and use his head to get to the top. Incidentally, the most natural mesomorph I ever knew was Chet Yorton, a Mr. Universe winner from back in the late sixties.

19

3

THE TRAINING LOG

THE SILENT REMINDER

The beauty of a training log is that it allows you to compete with yourself from one day to another, or one year to the next. Imagine being able to check back to the same week the previous year to find that you were in fact doing 10 reps with 300 pounds in the bench press, when currently you can manage only 4 reps with the same weight. Now your training diary is a "silent alarm," as Armand Tanny, a former Mr. USA and current writer for *Muscle & Fitness*, would put it.

A log makes you pay attention to what you do and how you feel during a workout. It also can bring a smile (or a frown) to your face when you look back over the years. The first thing you should write on the front page of your log is *Ceiling Unlimited*.

If you don't record the details of your training, then they will be lost forever. It's only the odd memories that you will be able to recall. Always record every set (except warm-up sets) and every rep. Note the poundages used and how the workout felt. Were you full of pep? Did a particular set seem easy, or did you feel drained during most of the workout? When you

Record-breaking training is a must for bodybuilding progress to continue. That doesn't mean you should constantly lift more and more weight at the expense of exercise style. Far from it. You do, however, have to fatigue the muscles on a regular basis to a greater degree so that they rebel by overcompensating. Overcompensation means added mass.

Many bodybuilders keep a training log or diary to keep track of exactly what they have done in the past and what they should aim for in the future. The first champion to keep such a log was three-times Mr. Olympia Frank Zane. He made a science of recording every set and rep and the emotions he had at the gym. His attention to detail paid off, big time. Dorian Yates, that multi-Mr. Olympia renowned for his physical mass, has kept a training log since he first started training. Without a training log, there is no way you can recall the exact weight, sets, and reps you started with at the beginning of your training. Yates can go back as far as 20 years with a few flicks of the pages. That's pretty handy information to have at your fingertips. Dorian doesn't go all-out on each set every workout. He deliberately cycles his training with specific phases of work intensity. But even on a "down" cycle, Dorian may attempt to break a record of some type.

(Left) Aaron Maddron
(Above) Dorian Yates
(Right) Aaron Baker

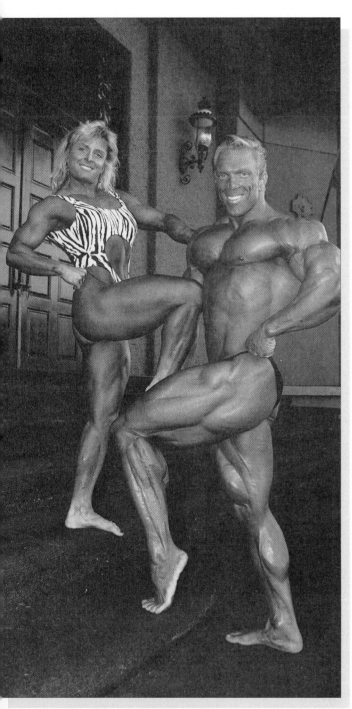

Sue Price and Dave Fisher

achieve a new record in an exercise, put a star (★) beside the set in your log. Those stars will indicate your progress. Record any injuries and the exact details of how they occurred. Were you trying a new exercise with too heavy a weight?

What about your food intake? This, too, should be recorded—not to the last morsel, but in its essentials. Pay particular attention to recording whether you are supplementing your diet with vitamins, desiccated liver, proteins, minerals, and so forth. This way, you will be able to correlate your workout capability and efficiency with your nutritional intake. You will be able to gauge the usefulness of various nutritional supplements. Mention the supplements by name in your training journal. Maybe you will find that EAS is better than MuscleTech or vice versa. Only carefully recording exact information will give you the data.

If the protein supplement you are taking is giving you constipation or diarrhea, say so. If you missed doing those all-important squats, record it. You missed the workout altogether to go shopping with your girlfriend? Date the page but leave it blank.

Measurements, too, should be recorded. The tape measure doesn't mean everything, but if statistics are recorded along with your fat percentage, then you have some worthwhile information.

Whenever you have photographs taken, paste them into your log. Don't include only your best pictures. Sometimes the unflattering photos can help you more than a fluke shot that shows you looking like the uncrowned Mr. Olympia. After all, the training log is designed as a training aid, not as a showpiece for the rest of the world. Any time you find yourself putting in remarks that are exaggerations or merely wishful thinking, forget about it! You are only writing yourself a valentine—a useless tribute to your ego.

Each day's entry must be dated. Note the time of day at which you train, and your style of training (cheat or strict). You also should record whether you are going to failure on your exercise, and whether you are performing forced reps, negatives, and so forth. If you are trying to shock a particular muscle into growth, then make sure that you record all the details and the results. How are the muscles reacting? Do you get a good pump?

Remember that bodybuilding is not the way it used to be. There are myriad training principles, although we are still a long way from knowing all the techniques that can improve muscles. Routines, like records, are meant to be broken. With a training log, the continual upward adjustments will help you improve on a regular basis.

A training log also will help you pace your workouts. When you keep track of everything, you become more aware that there is a tomorrow. You will be less inclined to go crazy for a new record at any particular session when you remember that you will have to do it again. It's far better for your record book to depict steady progress than an erratic pattern of ups and downs. In other words, it is better to beat your previous "best" in small jumps rather than to go for major advances in weight progression that will make steady progress in the long run more difficult.

Also, record your training goals. Define your needs and desires. Write down your short-term goals and your long-term dreams. When you see them in writing, they become more real.

You may wonder whether you should record your sets and reps at the conclusion of each set or at the end of the workout. This is entirely up to you. Some bodybuilders have no problem remembering their entire workout, whereas others have to write down the statistics at the conclusion of each set. Your concentration will not be broken if you write down the number of reps you performed with a certain poundage after each set. What's important is to keep a log, regardless of whether you make your entries during or after your workout.

Most bodybuilders train in a free-style, unplanned, random manner, which is not conducive to progress. In order to reach your highest potential in bodybuilding, it is necessary to train as efficiently and effectively as you can. A training log will help you maximize your efforts.

Milos Sarcev, Mike Matarazzo, and Mauro Sarni

4
MACHINES OR WEIGHTS?

WHICH ARE BETTER?

(Above) Jay Cutler works his chest on the Pec-Dek machine.
(Left) Ronnie Coleman displays an impressive "most muscular."

Are machines superior to free weights for building muscle? The debate rages on. Yet, one thing is certain: machines are getting more sophisticated day by day, and I suppose the time will come when machines will take over completely.

Joe Weider has put forth an interesting theory. He says: "I have often felt that the more aggressive person would, if left to his own devices, gravitate toward weights instead of machines." Ultimately, this theory was tested by Dr. Warren Chaney at the University of Houston.

A random selection of a hundred new gym members was divided into two groups of 50. One group was given instructions on exercises with machines only. The other group was instructed only in the use of free weights. None of these individuals had ever participated in a program of formal exercise.

Both groups then trained consistently for a month, one group using machines, the other a variety of free weights. After that month, each group cross-trained for a further month. That is, the free-weight users had to

work out on machines and the machine-trained people had to use free weights.

After those two months, the subjects were told they now could structure their own programs and use any equipment they liked: either weights or machines, or a combination of both. Although they didn't know it at the time, their choices were being recorded. Then Dr. Chaney matched the personality types with the equipment selected, and it turned out that Joe Weider was right. The more aggressive subjects selected the weights, whereas the less aggressive, almost without exception, chose to work out only with the machines. Dr. Chaney also noted that the individuals with the highest IQs tended to choose a combination of weights and machines.

Machines play an important role today in every bodybuilder's routine, but those machines usually are the old standbys. In rough order of popularity, these machines are the lat machine, thigh-extension and thigh-curl apparatus, and hack machines. Those machines have all been around for 35 years or more.

There is absolutely no doubt that the regular squat has a greater effect on building size and strength in the quads than any leg machine yet devised. Nor can one easily dispute that the wide-grip chin is more demanding than its equivalent machine, the lat-pulldown apparatus. But these machines do have their advantages. For example, the leg machines allow for multiangle "attacks." And the hack machine, the leg-extension apparatus, and the leg-press machine do, in fact, "hit" the legs from different angles. They don't rival the "king squat," but they do have a value of their own, because they bring something unique to the angle that cannot be obtained from squatting. At present, however, the leg machines should be viewed as an adjunct to squatting, not as a substitute for it.

Getting back to the wide-grip chin, it is the "king" of back exercises. At the least, it is on a par with the barbell-row movement. Wide-grip chins (behind the neck) give you a "feel" that no machine can duplicate fully. But it is also true that the lat machine allows for a fuller range of motion, and you certainly can do more repetitions if you choose. Here again, a machine can bring something extra to a workout.

An ideal combination would be performing a few sets of free-weight squats, followed by a few sets of hack-machine exercises or leg extensions. Likewise, chinning or rowing is frequently followed up with lighter lat-machine work. (The angle of pull can be changed just by altering the position of your body in relation to the pulley.)

The proliferation of exercise machines on the market has been staggering. They employ a variety of devices, from heavy rubber bands, springs, redirectional pulleys, and cams, to air pressure, levers, and even water. Many have valuable uses, especially in conjunction with basic barbell and dumbbell movements.

When Arthur Jones first brought out his line of Nautilus equipment, the bodybuilders of the world got extremely excited. The propaganda for Nautilus in *Iron Man* magazine in the late 1960s was enormous, and, I might add, seemed convincing.

When Jones writes, his passion comes through. He is one of the most inspiring writers on exercise whom I have ever read. But Arthur made two mistakes. He claimed that Nautilus training would produce top bodybuilders in a fraction of the time taken previously, and he attacked the modern disc-loading barbell, calling it an obsolete, injury-causing antique. However, there has not been one single Mr. Universe or Mr. Olympia who built his physique exclusively with these Nautilus machines. Every one of them used barbells and dumbbells extensively.

Nevertheless, it should be said that many Nautilus machines have some effects that cannot be duplicated by using weights alone. And Nautilus is not finished yet. They are still pushing ahead with ideas for new machines. Only the future will reveal what gems are in store for us.

As an example of the efficacy of machines, consider the following. For a long time a combination of bench presses and flying exercises has been the proven way to big pecs. So why, you might ask, did someone invent the Pek-Dek? Well, maybe the Pek-Dek was not exactly needed, but there are several reasons to justify its existence. The Pek-Dek allows for a totally isolated pectoral movement, which means you can work the pectorals in an upright position without the added stress of exercising the triceps. It is absolutely ideal for employing partner-assisted forced reps and "negatives," and with it there are none of the balancing problems associated with free weights. The Pek-Dek allows the "pre-exhaust principle" (more about this later) to be used in a unique way. In addition, the Pek-Dek works the chest with less of an energy drain and in greater comfort.

So, you see, the modest Pek-Dek has its uses. That doesn't displace the bench press as the chief builder of chests. But the Pek-Dek is of value even if it is used only as added variety. Remember the old "shock 'em" principle to maintain regular muscle growth? The Pek-Dek and many other machines as adjuncts can fulfill this purpose.

George Turner, owner of George Turner Gyms in St. Louis, Missouri, readily admits that machines do have worth, and he has scores of them in his various gyms. However, Turner points out in *Muscle & Fitness* magazine that these machines invariably do not take into account anything other than the height of the user. "They don't consider other variables," he says, "such as length of the thigh relative to the lower leg. There are so many incipient variables, like height and length and thickness of body parts, that there is no humanly possible way all these variables can be practically incorporated in the structure of the machine." As an example, Turner cites the thigh-extension machine (cam, pulley, or weight type), where the pivot point is the knee. With this machine, the force of the lever is displaced off the joint. This is in contrast to the natural movement of the squat.

Bruce Patterson works with intensity-plus.

Squats build size—period. You cannot expect comparative size gains from using machines. The thigh extension has its uses, especially for patients recovering from cartilage surgery or in other knee-rehabilitation cases. It serves the bodybuilder, too, but he must not expect great size gains from thigh extensions alone. It just doesn't happen that way, except for those lucky people who have genetically superior leg mass. For them, almost any exercise will build mountains of muscle. Moreover, the use of thigh extensions to failure is conducive to tendinitis, especially with machines where the movement starts when the lower leg is at an angle of less than 90 degrees. In fact, if any movement with the pivot point at a tendinous attachment is done to failure, it will invariably cause tendinitis.

Let's now consider the triceps. One of the most popular triceps movements is the triceps pressdown on a lat machine. It is used regularly by virtually every top bodybuilder I know, yet, ironically, it is not a size builder. Granted, it is a pleasant exercise, and it can sure rustle up a pump, but it does not help to stretch the tape at the end of the day. It's far better to use either the close-grip bench press or

the parallel-bar dip (both exercises work the belly of the triceps) than to waste your time with Nautilus-machine pressouts or pulley pressdowns.

You also will find that upright rows and overhead presses will do far more for your shoulders than cam- or pulley-operated machines. Nor will crossover cables and Pek-Deks give you the mass achievable with all forms of dumbbell and barbell bench presses and flyes.

Still, machines are probably destined to increase in variety and popularity, and I suspect many will achieve a degree of greatness. Yet, when it comes to building basic body mass, I can't imagine any machine, however well designed, that will improve upon the basic, natural disc-loading barbell and dumbbell exercises.

What I think will (or should) happen in the sport of bodybuilding is that the inventors will get busy designing platforms and benches that will allow us to use an even greater variety of free-weight movements. We cannot change the straight-line one-directional resistance path of the barbell, because gravity pulls downward only. But we can change the body position in all kinds of ways so that the barbell

The late Andreas Munzer

or dumbbell can work the muscles from different angles.

One of the first units to achieve this was the flat bench, invented 65 years ago. This was a great development, believe it or not. Prior to that, the barbell men were pressing from a prone position on the floor. It wasn't until the forties that the incline bench came into use. Then we got moon benches, preacher benches, high flat benches, decline benches, and so on—all designed to change the angle while you are using free weights. Still, there is room for more invention along these lines.

When the disc-loading barbell was invented, it was soon evident to strength-and-fitness buffs that this was a fantastic innovation. Almost overnight, it produced conspicuous changes in those who used it. Despite all claims, no machine ever made such an impact across the board. In fact, no machine or line of machines has been proven to build bigger muscles than free weights.

Nautilus developed the first machine to remedy the barbell's so-called disadvantage of only working the weakest part of a muscle (because only one part of the curl is difficult, but before and after that, it does not really tax the biceps). Since then, numerous other cam-operated machines have appeared on the market. The idea behind the cam, masterminded by Arthur Jones of Nautilus, is to provide for automatically varying resistance that corresponds to the strength curve of the muscle through its full range of motion.

When using a Nautilus machine, it is important to train in a slow, deliberate manner, so as to obviate any buildup of momentum. If you do try to "beat the clock" on a Nautilus, you also will be robbing yourself of its special benefit. In effect, the machine will become like a barbell, for you will find that your muscles will be taxed at one particular sticking point, and you'll be back to Square One!

I believe that all beginning bodybuilders should use barbells and dumbbells—that's the way to build a solid foundation. A novice will develop a feel for supporting and balancing free weights. He will enjoy the competitive aspect of lifting more and more weight. In addition, he will benefit from using compound movements. Weights allow for more muscles to be brought into play (with the accompanying neuromuscular coordination) than machines. Most guided-resistance machines work the muscles in isolation.

As a bodybuilder enters the intermediate stage, machines inevitably will play a role in his training—maybe an ever-increasing role. Machines can be used to hit specific parts, and that can improve proportion, balance, and symmetry. Also, machines often can be very effective in conjunction with weights.

To sum up, let me advise you to forget the argument as to which is better, machines or free weights. Take advantage of the benefits each has to offer. At present, free weights are definitely the fastest aids to building muscle mass. Machines can help to etch in the quality. Of course, neither weights nor machines will be effective unless you work with them. Only total dedication and regular workouts will produce maximum results.

The amazing back of Aaron Baker

5

REPS UNDER SCRUTINY

CHALLENGING MUSCLE-FIBER CONTRACTION

(Above) Chris Cormier works the triceps.
(Left) Serge Nubret

Bodybuilding is all about applying resistance to the muscles. By definition, a "rep" (or repetition) means more than one. The rep is what bodybuilding is all about. It is the heart of your workout. As Mike Mentzer says, "The workout itself is the sum of all sets, but the individual rep is the basis upon which the whole is built."

For a long time now, inventive bodybuilders have been striving to maximize the effectiveness of the core repetitions that make up the bulk of our training. Bodybuilders aspiring to added muscle size must understand that an individual muscle fiber (and there are millions of muscle fibers in our bodies) either contracts completely or does not contract at all. They should therefore design their exercises so as to contract as many fibers as humanly possible in the course of a set.

The goal, then, is defined, and there is no shortage of principles and techniques to involve the thousands of fibers that make up the individual muscle groups.

Straight-Set Reps

This is a basic system, and it's still the one that is used the most by beginning, intermediate, and advanced bodybuilders. It is the most effective method of training known. A set is a series of repetitions (usually 6–12), and more and more muscle fibers are involved with each repetition until the set is concluded.

Most bodybuilders perform 3 to 6 sets of an exercise when following this straight-sets system. However, total beginners are advised to perform just one set until they get used to the exercises.

Cheat Reps

The word "cheat" implies that you are doing something wrong, yet cheating (also known as "loose style") can be very useful if practiced correctly. Generally, you should not start to cheat during an exercise until you have performed the last rep possible in strict style. And then, of course, the less cheating, the better the effect.

When cheating, you should never "snap" an exercise through its range; rather, use a gentle body motion to aid the muscles in raising the weight.

Forced Reps

You will need a training partner for this method. When you are unable to complete a repetition using your own power, solicit the aid of your partner. He should then place his fingers under the bar and exert just enough pressure to allow you to make the lift. It is not advised to use more than two forced reps, and forced reps should never be used for every set in every workout.

Peak Contraction

In a peak contraction exercise, the muscle is under its greatest stress from the resistance (of a weight or a pulley apparatus) at the conclusion of a repetition.

The regular barbell curl is *not* a peak contraction exercise. The barbell passes through the most difficult part of the movement when the forearms are parallel to the floor, and there is definite relief as the bar approaches the shoulder area. The same type of relief is felt when you bench-press or squat. As the movement is concluded, there is very little resistance.

Peak contraction movements include wide-grip chins, spider bench curls, wide-grip rows, triceps kickbacks, standing leg curls, the crunch sit-up, gravity boot sit-ups, and leg extensions. Numerous machines also provide peak contraction exercises.

Actually, with some thought and rearranging of benches, pulleys, weights, and so forth, you can concoct dozens of new peak contraction movements. In fact, you don't even have to go that far. You can simply stop each repetition at the point at which you feel the most tension (this is called a partial or half rep).

Strict Reps

There is no doubt that strict reps have their place in a muscle-building routine. Few bodybuilders keep to strict form exclusively, but it is safe to say that most successful ones use fairly strict repetitions . . . most of the time.

When you perform exercises in good strict form, you eliminate the help gained by bouncing a barbell or swinging the body to raise the weight. In other words, you make the muscles themselves do all the work.

Starting the movement slowly is of prime importance when you exercise strictly. Do not kick out at the beginning of a leg extension. Likewise, when you are curling a barbell or dumbbells, begin the lift slowly and deliberately. When you are pressing, begin methodically. No back bend or jerking! Never bounce out of a squat, for that will only wreck your knees. No bouncing! When you are doing calf raises, use deliberate up-and-down motions with a full stretch.

Rest-Pause Reps

This method has been used ever since barbells were invented. Rest-pause is not a system to be followed all the time, but it does permit you to make gains in tendon and muscle strength and in overall size in a few weeks.

It's a simple technique. After warming up for a particular exercise, you load up the barbell sufficiently to allow just one repetition. Assume you are bench-pressing, press out one difficult rep and replace the bar on the stands. Let 10 to 20 seconds elapse, and then perform another repetition. After a similarly brief rest, perform yet another rep, and so on. Each time you allow your body to partially recuperate. As the reps mount, you may have to reduce the weight slightly to attain 6 or 8 reps.

Rest-pause is a very strenuous way of training, and it is not for extended use. It's best employed only infrequently to break a sticking point.

Nubret Pro-Rep Method

Serge Nubret of France has a unique way of training. At one time he trained very heavily and could curl more than 240 pounds and bench-press 500. Today he exercises using a singular, seldom-practiced principle. Nubret introduces progression into his training, not by constantly pushing the poundages higher and higher, but by pushing his rep count up. In actual fact, it is a double progression, as he also tries to "race the clock."

If, for example, it took him 45 minutes to do 30 sets of an exercise one day, the next workout day Nubret will try to squeeze out 31 or 32 sets in the same length of time. He often wills himself to beat his rep record of the prior set. Using the seated dumbbell curl, Serge may start off by doing a set of 10 reps with a 45-pound dumbbell; in his second set, he will do 11 reps, his third set 12 reps, and so on, all with the same weight. The secret of Nubret's method lies in his ability to "feel" an exercise by focusing all his concentration on the movement he is doing at the time.

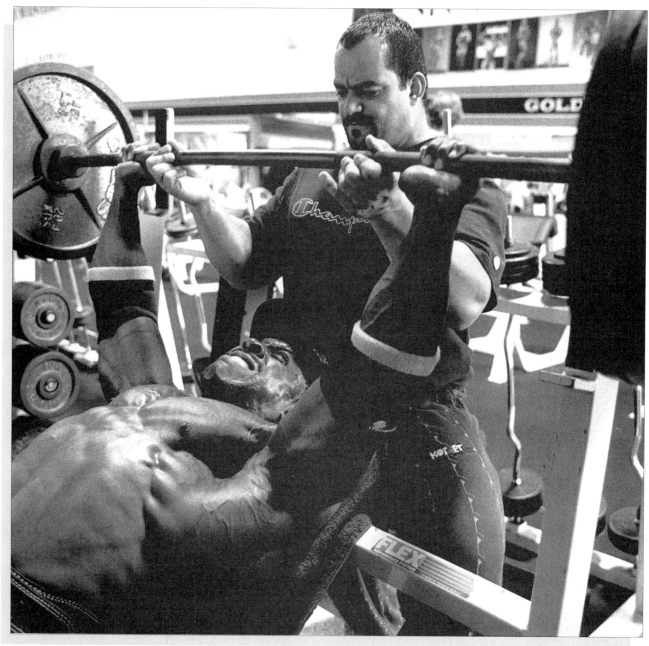

Melvin Anthony gets a "spot" at Gold's Gym, Venice.

Superset Reps

All muscles are actually pulling muscles. They can only contract and shorten, and therefore they cannot push. The upper arm's biceps muscles, for instance, contract and pull the forearm upward. The opposite movement (straightening the arm) involves the triceps muscles at the back of the arm, and they pull the arm straight. Nevertheless, bodybuilders refer to pulling muscles and pushing muscles. Exercises in the pushing category are the standing press, supine bench presses, push-ups, triceps extensions, and leg presses. Those often referred to as pulling exercises are upright rows, curls, chins, bent-over rowing, and thigh curls.

The original concept behind supersets was to alternate, rapidly and without rest, two exercises: one pulling and the other pushing. The most common combination was to alternate curls with triceps extensions. However, many champion bodybuilders simply would alternate two curling movements or two pectoral move-

Kevin Levrone *(left)* and Shawn Ray *(right)* take the spotlight at the Olympia.

ments or two triceps movements, not caring whether a particular muscle was worked alternately with its antagonistic partner (for example, the biceps and triceps).

Today, the term "superset" merely indicates the alternating of two exercises in rapid succession. A few weeks on this type of exercising routine can jolt the muscles into new growth. It is a severe form of working the muscles. Too much of it could cause you to grow stale and bring you to a standstill. Paradoxically, a standstill can be broken with a week or two of supersets.

Compound-Training Reps

Compound training, sometimes known as "giant" sets, is definitely an advanced technique of muscle building. A compound set for the deltoids would involve performing three or four shoulder exercises, one after the other, with a minimum of rest between exercises. An entire shoulder routine using the compound-training principle could look something like this:

Press-behind-neck	10 reps
Seated dumbbell presses	10 reps
Upright rowing	10 reps
Standing lateral raise	10 reps
Short rest	10 reps

You repeat the entire routine twice, for a total of 3 sets.

Pre-exhaust Reps

I invented this technique around 1964 and documented my findings in *Iron Man* magazine in 1968. Arthur Jones then incorporated the technique into his line of Nautilus machines in the early seventies. Pre-exhaust is the battering of a specific muscle with a carefully chosen isolated exercise, immediately followed by a combination movement. It is currently a great favorite with many top bodybuilders.

Let's use the chest as an example. As you may know, the triceps are involved in many of the recognized chest exercises, and, for most people, they are the weak link. That is, when you do dips, bench presses, or incline presses, the triceps are worked hard and the pectorals only moderately. This means that your triceps will grow more rapidly than your chest. That's fine if you already have a big chest, but if you want to develop your pecs, the best way is the pre-exhaust method. Here's how to do it.

To get around the "weak link" triceps, isolate the pecs first with an exercise like the dumbbell flyes, where the triceps are not involved directly. After a hard set, carrying the exercises to the point of failure, proceed immediately to the second exercise, such as incline or bench presses. When you do the presses, the triceps will temporarily be stronger than the pectorals, which are in a state of near exhaustion from the first isolation exercise. You are not limited by the weak link in the triceps.

Sample All-Around Pre-exhaust Schedule

Shoulders
Lateral raises (isolation movement)
Press-behind-neck (combination movement)

Chest
Incline flyes (isolation movement)
Incline bench press (combination movement)

Thighs
Leg extension (isolation movement)
Full squat (combination movement)

Back
Wide-grip chin-behind-neck (isolation movement)
Bent-over rowing (combination movement)

Biceps
Scott/preacher-bench curls (isolation movement)
Narrow-grip chinning the bar (combination movement)

Triceps
Triceps pressdowns (isolation movement)
Narrow-grip triceps bench press (combination movement)

Calves
Standing calf raise (isolation movement)
Jumping rope (combination movement)

Pyramid-Training Reps

This method is used very widely because it allows the bodybuilder to start easily, build up to a peak, and taper off effectively.

You start with a set of high reps (12–15) to warm up the muscles. In the following set, some weight is added and the reps are diminished. You do this with each set until only a few reps are possible. Then it is time to come down the other side of the pyramid. With each successive set, your weight load is decreased to allow for the extra repetitions. A typical pyramid routine for the bench press might look like this:

Set 1:	20 reps,	120 lbs
Set 2:	10 reps,	150 lbs
Set 3:	8 reps,	170 lbs
Set 4:	6 reps,	190 lbs
Set 5:	6 reps,	210 lbs
Set 6:	3 reps,	230 lbs
Set 7:	8 reps,	140 lbs
Set 8:	12 reps,	120 lbs

6

INJURIES

HOW
TO AVOID
THEM

M ost sports at the competitive level are dangerous, and virtually all athletes endure the constant risk of injury. The worst part for a bodybuilder is not the pain of an injury (although that can be unsettling), but the annoying inconvenience of not being able to train as he wishes. Even a small muscle strain can keep you away from an all-important exercise for many months.

Weight trainers can incur injuries in the form of tendinitis, muscle tears, strains, sprains, or even bursitis or hernias. However, it's important to note that the primary cause of injury to bodybuilders has more to do with carelessness than anything else. With common sense and care, there is no reason why you shouldn't enjoy a successful bodybuilding career totally free of injury. Yet, in reality, all weight men incur at least minor injuries, in part probably due to the show-off aspect of the sport.

A few years ago, pointing jokingly to a 150-pound dumbbell at my Toronto exercise store, I said to my friend Johnny Fitness, "Remember in the old days when we could lift that weight with one hand?" Johnny walked over to the dumbbell and, cold turkey, cleaned it to his shoulder. He then jerk-pressed it overhead a couple of times and lowered it to the floor. I smiled. "I know I couldn't do that," I said. "Boy, you're still as strong as a bull." Johnny's grin of satisfaction turned ice cold as a searing pain ran from his neck to his upper arm. The nerve injury stayed with him for three years and only recently has shown signs of repair.

Never be tempted to pick up a heavy dumbbell in order to heave it overhead. This can lead to a pinched nerve, which could cause you pain for years, with an almost immediate loss in muscle size and strength in one arm.

"Never begin your workout with barbell curls," warns Franco Columbu, and he should know, as he is a registered chiropractor. "The exercise itself is good, but do not start your training with this movement. Curls 'lock' the elbow joint, leaving the biceps vulnerable to injury."

Warming up is an integral part of training. Never neglect it. Before you go into any exercise, you should do at least one set of 10 to 15 repetitions with about 50 percent of your limit for that number of reps. If you don't like the idea of performing a high-rep warm-up,

(Above right) Craig Titus, Gold's Gym, Venice (Left) Franco Columbu works his shoulders with front raises.

then do 2 to 3 sets of lower-rep warm-ups, again using about 50 percent of your maximum.

Another way of inviting injury is to neglect the importance of keeping in the groove. Let me explain. As we train, each of us develops a groove, or line, in which the weight travels. Take the bench press. The weight is invariably lowered to the nipple area of the chest. Year in and year out, you bench-press in the same way. Your pecs build up. Your strength triples. But one day, while happily benching away with your regular poundage, you get the idea that maybe you should lower the weight to your upper pectoral. It seems a good idea . . . then . . . wham . . . ouch! Something tears. Hot needles in your chest. You've done it: a muscle tear! Why? Because you put all the stress from the pectoralis major onto the pectoralis minor, an area that just couldn't cope with that kind of resistance. You hit a different groove.

When you get the idea that you want to change an angle or perform a new exercise, even if that change is minimal, you must approach it like a beginner. The bodybuilder who can bench-press 300 must not change the groove—at least, not unless he is prepared to use only one-third of the weight. Real strength will come quickly, but you can't force it without risking injury.

Michael Francois of Ohio

The same goes for any other change. You cannot suddenly do heavy incline presses if you have never done them before, even though you can bench-press 500. If you want to forge a new groove, then begin with a light resistance and use a regular and unhurried progression.

You must take particular care in certain key exercises. With the squat, for example, always keep your back flat and your head up, and lower slowly to the thigh-parallel-to-floor position. Never bounce out of a squat. Deadlifts also should be performed with a flat back, your knees bent, and your head up. Do not rebound the weight from the floor.

When performing bent-over rowing, keep your knees slightly bent and your back flat. Actually, this is a dangerous exercise, because the lower back is terribly susceptible to strain when you use very heavy weights. I would prefer you stay with T-bar rowing, or, better still, single-arm dumbbell rows, where one arm can support your torso on a bench.

You wouldn't believe the number of bodybuilders who suffer shoulder tears. Because of their three-headed and complex formation, the deltoid muscles can become injured easily. Such injuries don't result from barbell presses as often as from the precarious lateral-raise movements. The reason behind this is that when you raise a dumbbell (elbows locked or almost locked) at arms' length, there are scores of grooves, or pathways, in which it can travel, and if you hit an unused groove with a weight that is too heavy, a tear may result.

Another fiendish exercise, which when correctly performed is little short of magnificent, is the preacher curl carried out with either a barbell or dumbbells. The danger increases as the arms straighten, especially if the angle of the preacher top is shallow. Never, never, never allow the weight to bounce upward from this straight-arm position. Physique star Dave Spector did that, and it was Operation City for him as a result.

You also should beware of alternating certain exercises. You can alternate biceps and triceps exercises, for example, or chest and back. However, take particular care not to alternate chest and shoulders exercises. Because of the muscle arrangement of the chest and shoulders, if one is exercised after the other, it is likely to lead to injury.

Tendinitis, the inflammation of a tendon, is another form of injury that can attack tennis players, wrist-wrestling champs, and house painters, as well as bodybuilders. Tendinitis can result from exercises that place a strategic joint in a vulnerable position. The single-arm triceps stretch, when performed with heavy weights, is a prime producer of this painful condition.

It is sound policy to leave out any exercise that you suspect of causing joint or tendon problems. Either that, or limit the poundage used in such exercises and take special care to concentrate on good exercise form.

Unquestionably, you are more prone to injury when your concentration lapses. So, keep your mind on what you are doing. Do not listen to others while training. Pay attention to only what you are doing. The chatterboxes in the gym are wasting their own time. Don't let them waste yours, as well.

What is pain? Pain is a kind of defense mechanism of the body warning you that injury is occurring or has occurred. The pain of lactic acid buildup in the muscle is acceptable as you push your reps to the limit. But sharp, searing, hot-needle pain is different. It means a real injury is occurring and you must immediately stop what you are doing.

Should you become injured, try to determine its severity. Apply ice cubes directly over the area of pain as soon as possible. This will help to reduce the swelling and the inflammation. If the injury prevents you from moving your limbs in any direction, see your doctor. Competent professional help is always best, for even the smallest injuries can worsen, and in some cases they might become lifelong burdens.

Many bodybuilders who have injured themselves manage to "work around" the injury. That is to say, they continue to train but only use exercises that don't aggravate their condition. On no account should you perform any exercise that awakens the painfulness of an injury.

In most cases, a torn muscle, a strain, or a sprain requires rest. Subsequent training must, therefore, not involve the affected area until it has healed. Then with caution, resume regular training, always mindful of the mistake that caused the injury in the first place.

One helpful safeguard against injury, though no guarantee, is to practice some form of stretching exercise prior to beginning your workout. To get the fullest stretch, a muscle must be warm. Therefore, in cold temperatures it's a good idea to wear a sweat suit. The proper method of stretching is to take your time and allow your muscle to lengthen gradually as you bend and reach (no bouncing or jerking). In all stretches in which you are attempting a maximum effort or bend, try to hold the point of complete extension for a full 15 seconds.

Preparation for Lower-Body Stretching

Lean the body over, as depicted. Slowly . . . now. No bouncing. Hold the position for around 30 seconds, allowing your body weight to stretch the backs of your legs. You will feel some mild discomfort. As you get used to this position, it will benefit you to force it slightly. Try for a lower forward position.

Bent-Torso Pulls

As shown, lean over to one side, and try to pull the upper body down until the chest touches the thigh, using your arms. Keep your feet at least a yard apart, and twist the waist into the direction that you are stretching without bending at the knee. Work each side 10 times, and hold for about 10 seconds each time. This is an excellent movement for flexibility for the entire back, the hamstrings, and the calves.

Floor Stretches

Spread your legs outward as far as you can. Working one leg at a time, stretch into the knee area with your head down. Try to pull your chest down to the thigh with the use of your arms. Hold the stretch for 10 seconds. Stretch each leg 8 times.

Standing Groin Stretch

Standing with feet apart, pull down on each leg. If you wish to maximize the stretch, try to touch your head to your toes. Remember, no bouncing! Just pull the body down slowly to the toes and hold for 10 seconds. Try the exercise 6 times.

Back Stretch

Roll up your shoulders, supporting the body with your upper arms on the ground. Gradually and under control, lower your legs behind your head. Try to touch the knees to the floor by the sides of your head. Hold for 10 to 20 seconds, or longer when you get used to the position. This is very good for maintaining a limber spine, and you need to perform it only once.

7

CYCLE TRAINING

PUSHING TO A PEAK

There's an old saying that you are sure to resist: "Make haste slowly." Ironically, if you push yourself in your workouts each and every session, you will drive yourself to a sticking point. There is a difference between going all-out in your training every session and cycling your efforts to complement your physical condition. With cycling, you build intensity up to a peak, and then you taper off and consolidate your gains by demanding somewhat less from your sessions. All athletes do this. It's a kind of controlled progress.

Any condition that you can maintain year-in and year-out is not a peak condition. Your body can be pushed to peak condition, but then it must rest. The edge will go inevitably, but it can be recalled and superseded with a new thrust. "Making haste slowly" applies very much to bodybuilding. In fact, it is the fastest way of building good-quality muscles.

John Cardillo, one of Canada's top amateur bodybuilders and a successful gym owner, is not particularly enamored with the idea of cycle training. After a layoff, Cardillo likes to get back into shape quickly and drive at full speed, going all-out each workout. He believes that his body will tell him if he is overtraining, and that at such times he will rectify the situation by taking a few days off.

There are actually a large number of bodybuilders who train this way very successfully. My own observation is that they are young and very enthusiastic and just do not burn out. Cycle training is usually practiced by the older or more experienced bodybuilder who may not have an all-consuming, fanatical approach to his workouts. Day after day of blitzing his muscles simply doesn't fit his temperament. I would say that most, if not all, professional bodybuilders practice cycle training. Conversely, the majority of amateurs do not.

Top bodybuilder Clarence Bass puts the case for cycling your training this way: "Gains in muscular size and strength can only be forced temporarily. Long-term gains must be coaxed, induced in an agreeable manner, by gentle persuasion. Few bodybuilders are willing or able to strain to the limit continually. I doubt that anyone really wants to do curls, or any other exercise, until they are blue in the face—not on a regular basis, anyway. The mind rebels. It will not face such effort day after day. Bodybuilding progress, like progress in any other activity, is irregular; it's full of peaks, valleys, and plateaus. Don't expect to make continuous progress. A bodybuilder should push for a while, back off, and then push again."

(Opposite) **Ronnie Coleman goes for the biceps concentration curl.**
(Above) **The one and only, Arnold Schwarzenegger**

41

Joe and Ben Weider congratulate Dorian Yates on yet another Olympia win.

One thing everyone seems to agree upon is that you can't run your body at full throttle all the time. You either have to cycle your training intensity or take occasional layoffs.

Boyer Coe, who has trained numerous top men and women, feels that layoffs are counterproductive physiologically. According to Coe, "You're much better off taking a period of what the Soviet weightlifters call 'active rest.' During this time, I do bodybuilding workouts of lesser intensity. I also concentrate on building up my weakest body part."

Tom Platz has a similar view. In the past he trained so hard that he sustained a mind-boggling list of injuries. In his eagerness to reach the top, he strained to a point where he injured his shoulder, tore his biceps muscle, brought on varying degrees of joint stress, and actually burst blood vessels in both his eyes. Mercifully, he has recovered fully from these injuries, and now trains with a good deal less intensity. "In the old days," he says, "I didn't know how much was too much. I have learned from my experience. No more torn muscles and burst blood vessels. Violent, massive, reckless effort is not the answer." Today, Tom Platz trains more moderately, but he is still known as one of the world's hardest trainers, even at the age of around 40.

Some bodybuilders cycle their training from one day to another. In other words, they perform a light (less intense) workout every once in a while. With others, this method involves two heavy workouts and two light workouts every seven or eight days. Still others may choose to "go light" once every two weeks or so. The number of light workouts you allot yourself will depend on your metabolism, your rate of recuperation, and your tolerance for vigorous exercise.

The most common form of cycle training, however, is to gradually build up your training poundages, number of exercises, and duration of workouts to peak for a particular contest, and then to rest up by changing your exercise habits and downgrading your training intensity. Bodybuilders like Platz and Coe will stop all heavy training and increase their aerobic activities like bike riding, running, and swimming, all with one aim: to allow the body to regenerate. The comparative rest provides a solid base from which you can proceed to a new cycle of increasing intensity, for the next contest or photo session.

Cycle training is good for you. The body just cannot take a nonstop pounding. Even Mike Mentzer believes in cycling his training efforts.

Also, there is evidence to show that those who push their muscles to failure and beyond for long periods of time and without interruption may over-stimulate their adrenal glands. It's helpful to have a huge flow of adrenaline in an emergency in which you may have to run or fight for your life. When this response is triggered too frequently, however, the adrenal glands become overtaxed and exhausted, with a resulting reduction in output. In short, you become lethargic, lose interest, and show all the signs of what is known as the "overtraining syndrome." Research has shown that maximal stress cannot be endured for more than two or three weeks before this state of physical staleness sets in.

The answer, then, is to cycle your training to incorporate a steady, progressive buildup, but also to be aware of the dangers of exhausting the body's inner vitality. When you plan on forcing yourself to a new plateau, do not make an all-out effort that exceeds two or three weeks. Chances are, you will be defeating your aims if you do.

After a contest or other event for which you have peaked, you should wind down your training to a comfortable level with enjoyable exercises like

Boyer Coe

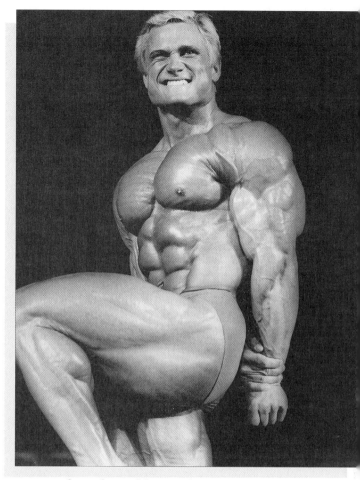

Tom Platz, the Golden Eagle

bike riding or swimming. Do not pig out on food to the extent of getting fat. Make a point of not gaining more than 10 to 15 pounds over your competition weight. If you allow your body to bulk up too much, you only make matters more difficult for yourself the next time around.

After winding down with aerobic exercise, do not force yourself right back into heavy training. Allow yourself time. Let the desire return to you naturally. And when it does, harness your enthusiasm and control your workouts. You will be surprised how you can make your workouts progress steadily, even when you don't drive yourself with all-out intensity.

Ultimately, of course, when your next contest approaches, you will push for a new peak, and because you have held back and paced your progress, a new peak of development will arrive assuredly. Cycle training is used by all athletes. It works for them, and it will work for you. It is one of the secrets of bodybuilding success.

8

POWER THINKING

MENTAL PROGRAMMING FOR SUCCESS

L ike it or not, computers are with us for good. Our bank has our financial status on computer, and the government has our personal statistics on computer. The government even has the audacity to give each of us a number—like a convicted criminal—and that number is on computer. Stores, credit card companies, libraries, licensing bureaus, tax offices, and who knows what other institutions, have information about every one of us on computer. Rightly or wrongly, many of us resent the computer age and the invasion of privacy we suffer as a result.

Yet, we too each possess a computer of our own: our brain. And our brain is far more complex and versatile than any billion-dollar computer system invented by man. The brain controls everything we do, and, like a computer, it can be programmed. We can program our brain through positive thinking, visualizations, and even enthusiasm.

Goals and Promises

The superachievers in bodybuilding today have programmed their brains for success. They have set themselves on a path to physical achievement that is not altered easily. They have set definite goals. They are totally confident of their ability to achieve them, and they persist relentlessly in their quest. Because they have made a promise to themselves, their goals are virtually cast in stone.

Have you ever noticed that before you perform a set of repetitions, as you lie down on the bench or take your hand spacing on the barbell, you actually talk to yourself? Some bodybuilders even talk aloud. What is it they are saying? What do you say before you attempt a set? You tell yourself how much effort you are going to allocate to the movement. You program your brain: "I'll do ten reps" . . . "I'll try six strict reps, two or three cheating" …"I'll go for five reps to warm up." If you didn't program yourself this way, do you know how many reps you would do? None!

Setting goals, both for the short term and the long term, is essential to success in bodybuilding. "When I'm not in training for a contest," says Franco Columbu, "I try to train heavy and hard. But my mind asks me why I am doing it if I am not getting ready to compete. So my body doesn't go all-out." Contests on all levels—

(Left) **The amazing legs and glutes of Hamdullah Aykutlu.** *(Above right)* **Gary Strydom, Venice, California**

novice, state, national, international—serve to give bodybuilders the incentive they need to make them train harder. Flex Wheeler feels similar to Franco. He says, "During the greater part of the year, when no contest is imminent, my training is not fueled with the intensity or energy I know I am capable of generating."

Visualization

It is one thing to believe in the power of the mind and quite another to harness that power. Visualization is one important technique for consciously bringing the power of the mind to bear on workout performance.

45

Charles Glass and Roland Kickinger

The way you visualize your performance before you actually attempt it will greatly influence the result. For example, if an Olympic lifter has genuine doubts that he can lift a heavy barbell, then there is little chance that he will achieve it. Conversely, when you have 100 percent belief in your ability to succeed, coupled with 100 percent effort, you can take yourself to the very limits. In fact, it is believed that visualization actually helps to develop the neural paths that are required for precise control of physical activity.

There is no doubt that incentives can so excite and inspire the mind that the effect spills over and produces a dramatic increase in physical effort and achievement. If your visualization or dream is vivid enough, the subconscious makes positive adjustments that clear the way and expedite the realization of your goal.

Arnold Schwarzenegger, the personification of success both in bodybuilding and in his show business career, has said, "When I am exercising my biceps, I see them as two huge mountains, filling up the room, getting bigger and bigger."

Inspiration

In his essay "The Energies of Man," psychologist and philosopher William James explains why some of us are able to perform at a higher level than others: "Either some unusual stimulus fills them with emotional excitement, or some idea of necessity induces them to make the extra effort of will. Excitement, ideas and effort are what carry us over the dam."

Mr. Universe Boyer Coe uses music to excite his urge to achieve. "Sometimes when I go into the gym to train," he says, "I am anything but inspired to go at it hard, but as a lover of music, I get some of my heavy music going . . . and my workouts quickly take on a new meaning."

Some bodybuilders find they get inspired by training with the opposite sex. In fact, numerous male bodybuilders make a point of selecting female training partners. This, they claim, provides them with the inspiration to bring their training efforts to a new high. Presumably, the women also derive inspiration from training alongside the men.

One famous bodybuilder, who desires to remain anonymous, admits: "I have more incentive to give my everything to my workouts when I train with a woman. It seems to stir up my hormones, or flood my system with cholesterol, so that I double my efforts."

When you really believe in yourself, or are inspired by some outside source, and are completely determined to achieve your goal, your mind releases the amount of energy you need to give it your best shot.

Concentration

It is necessary to give your total concentration to a particular set if you are to advance to a new plateau of physical development. But this is more difficult than you may imagine. Concentration means single-mindedness—holding one thing in your mind to the exclusion of everything else—and it has to be learned.

It has been said that the mind cannot concentrate easily on one thing for more than a few seconds, and if you can learn to focus your attention on one thought or object for 12 seconds or more, then you have the ability to concentrate fully.

By using this type of focus on your sets, excluding all outside distractions, you will be using your mind to maximize your progress.

If you ask any serious bodybuilder what got him interested in the sport, chances are he will tell you that he saw a well-built guy somewhere, or else he saw a picture of a bodybuilder on a magazine cover. Either way, he was inspired to take up bodybuilding himself. Sometimes it is this first inspiration that keeps an indi-

Pavol Jablonicky

vidual training intensely for a lifetime. In most cases, however, the real stimulator that keeps someone training is the sight of his own progress. When you see your muscles growing, when you witness an increase in muscle strength, definition, and density, and when friends start remarking about your development, you naturally strive to go for more.

Sometimes, however, this muscle madness deserts us. It may be our age, or perhaps some other interest has slowly taken precedence. Nonetheless, we still want muscles. It is at such times that we need to summon the power of our mind to program our brain for success. With enough mental effort, we can make ourselves into anything that our genetic endowment permits.

Many bodybuilders with potentially great physiques are held back by a lack of positive programming. You can meet a guy with enormous Mr. Universe potential, only to find that, mentally, he will never pull it off. He does not believe in himself because he is full of negative programming.

Decide what you want from bodybuilding. Paint that picture in your mind, and follow through with whatever it takes. Limitations develop as a result of limited thinking. You can train your mind as hard as you train your body, and in so doing, totally control your bodybuilding gains. Bring your muscular development to what you want it to be. Develop confidence, poise, and charisma—and come out a winner. It's your call.

9

SUPER-STRUCTURING YOUR ROUTINE

IMPROVING YOUR LEVEL OF EFFICIENCY

M ost of my day is spent working on my magazines, *MuscleMag International* (which has the second-greatest circulation of all bodybuilding magazines) and *Oxygen* (started in 1997 for women aspiring to fitness). I am on the phone to writers, photographers, fitness athletes, and bodybuilders for about 50 percent of my day. Another 20 percent is spent on organizing articles, photographs, and artwork, and on correspondence with my contributors.

But 30 percent of my day is spent in my muscle store in Mississauga, Ontario, and there, much of my time is taken up by discussion. You can guess what the topic is—bodybuilding, and more specifically, how to make the fastest progress possible.

Now, I myself have been in this game for 40 years. In addition to publishing two magazines, I have written scores of courses and dozens of books, including *Hardcore Bodybuilding*, which actually made the best-seller lists. I also invented the much used "pre-exhaust" technique.

Big deal! I'm still a schmuck. Why? Because I still can't tell you exactly how to train. All my experience, I often think, has taught me only one thing: that there is more than one way to skin a cat!

Every summer literally thousands of aspiring young bodybuilders and fitness athletes make the trip from all over the United States and Canada, some even from Europe as well as Asia and Africa, to the mecca of bodybuilding—Southern California. In their hordes, they visit Gold's and World's gyms in Venice, a beach community in Los Angeles County. Many have saved all year to make the pilgrimage of six, eight, 10 weeks, what have you, and they are all there for the same reason: to learn from the stars, to be in their proximity, and get the inside scoop on training. As trainer Vince Gironda says, "There are more books, magazines, and training establishments than ever before in the history of bodybuilding, yet everybody is looking for one thing: information. And they can't get enough of it."

When these keen young bodybuilders get to California, they are usually shocked out of their socks to see their favorite superstars using standard barbell and dumbbell exercises. Somehow, they ex-

(Above right) Craig Titus completes an intense set of incline presses.
(Left) Popular Lee Priest gives an impressive "most muscular."

pected to see something new and secret being performed, but Jay Cutler is over there doing barbell rowing, Paul Dillett is doing calf raises, Nasser El Sonbaty is doing barbell curls, Lou Ferrigno is bench-pressing, and Platz is squatting. Our budding bodybuilder can't believe his eyes. It is all so normal that it's shocking.

Ironically, in spite of all this normalcy, our muscle pilgrim is beset by confusion, because after a few weeks he is confronted with so many permutations of how often to train, how many sets and reps to use, what food supplements to take, and how much weight and intensity to use, that he just doesn't know what to believe. Chances are, if he asks six bodybuilders for the best way to train, he will get six different answers.

When looked at in its simplest way, bodybuilding is little more than making a muscle work to lift up a given weight. In fact, a beginner usually can make good progress even with very haphazard training. He may train only once every few weeks and then go through a period where he trains every day. He may do a hundred reps, or only 5. He may party

Abs galore are displayed in this competition shot of Kevin Levrone, Nasser El Sonbaty, and Shawn Ray.

instead of sleep; he may drink, smoke, and eat junk food. Still, gains will come. It shouldn't happen, but it does. So much for scientific bodybuilding.

But when you break all the rules, you will pay for it sooner or later. Your early progress will not continue because of one unalterable fact: the bigger you get, the harder it is to get any bigger. So, in the beginning even totally inefficient training will yield results, but in time the gains will slow down or even come to a halt. It is at this point that you will have to take stock of yourself. From now on, you must maximize your training efficiency. That means that each variable must be given attention. In short, you must do your best on all fronts.

Apart from diet, food supplementation, and relaxation, there are numerous variables involved in train-ing, such as the number of repetitions, number of sets, speed of training, order in which exercises are performed, amount of weight, number of workouts per week, amount of rest between sets, choice of equipment, number of exercises used, and strict or loose exercise style, as well as mental attitude.

Right now, chances are that your training, with its myriad variables, is not as efficient as it could be. Let's check it out, variable by variable.

Training Intensity

To keep a muscle growing, you have to keep increasing the punishment you are delivering to the muscle *on a regular basis.* This is enough of a reason for holding back in your workouts if you have just taken a break

from training because of holidays, school exams, a job, or what have you. After all, it would hardly make sense to jump right back into all-out super-strain workouts. With this kind of punishment, you would probably end up with sore and injured muscles. The correct way to go about getting back into training is to make a slow but planned effort to pace your training intensity progressively.

Progression

Only do today what you can supersede tomorrow. It would be far better to increase your bench press by 5 pounds this week, knowing that you can add another 5 pounds next week and another 5 pounds the week after that, than to go all-out now and fail progress later on.

You see, once you have lifted a weight heavy enough to stimulate your muscle fibers, lifting a heavier weight doesn't give you better results. That's why, when you go to Gold's, Powerhouse, or World's gyms, you often will see a champion bodybuilder using only light or moderate weights. He is at the low end of a cycle and beginning to increase intensity on a regular basis. Each workout, he will do a little more. Perhaps he will add another set, or maybe he will add a few pounds to the bar. If he is only handling 40-pound dumbbells, then that is all he needs at this time to stimulate growth. As he nears the time of a competition, he may well be using 60-pound dumbbells, but he will not have made the transition with just one or two jumps. He will have increased the poundages, reps, and sets progressively, one step at a time.

How Many Repetitions?

Theoretically, one could justify the single (all-out effort) rep system for building the largest muscle size. Some weightlifters are after explosive strength and use that system. But such a system does not serve the bodybuilder's needs. The competitive weightlifter who attempts to fire off as much muscle fiber as possible within the few seconds it takes to complete a single rep will not gain an overabundance of muscle tissue. For muscle-building purposes, it is far better to overload your muscles regularly with repeated sets of 8 to 12 reps, progressively increasing weight resistance whenever possible. This will give you the kind of body development you have always wanted.

Regularly performing plenty of sets and reps (volume training) is important, as it serves two purposes. It contributes to plumping up your muscle cells individually, and it helps you to build new capillaries. This is why trained bodybuilders get a much bigger pump than Olympic lifters. Most advanced bodybuilders can add well over an inch to their arms by doing a few dozen close-handed floor dips.

Another result of volume training that has come to light is that the glycogen stored in the muscle can be increased significantly. According to bodybuilding author and photographer Bill Dobbins, "Glycogen is carbohydrate energy stored in the muscles. For each gram of glycogen, the body will store 2.7 grams of water, all of which adds to muscle size and shape." This is why glycogen-starved bodybuilders who are on too low of a carbohydrate diet appear stringy and small. It also explains the usefulness of "carbing up" a few days before an important contest (often done by eating sweet potatoes).

Having given you the case for volume training, now let me say there is also a case to be made for employing power-building techniques (as used to prepare for power lifting) on an occasional basis. Using heavy weights for lower repetitions can act not only as a tonic, but can give the muscles a new dimension of experience. A change is as good as a rest. Chris Cormier, Lee Priest, and hundreds of other bodybuilders employ heavy training every now and then. It improves the strength of muscles, tendons, and ligaments, among other things, and that is useful for achieving higher repetitions and upgrading training poundages when more regular workouts are resumed.

Exercise Style

Watch Shawn Ray train. His curls are magic. There is a rhythmic flow, a cadence, that is beautiful to behold. For the most part, you should train with good exercise style, working your muscles over the fullest range possible. For instance, start your curls with straight arms; do not begin with your torso leaning forward, elbows bent, swinging the bar upward.

Cheating (loose-training style) is a sophisticated technique. You need to know how much to cheat and, more important, when. Arnold Schwarzenegger made a habit of training in a very strict style for the first 8 reps, and then he would cheat more and more as he labored through the last 4 reps. That way, he got the benefit of both styles, but only after he had exhausted the benefits of doing 8 quality repetitions in faultless exercise style.

Speed of Training

Increasing the speed of your training—reducing the rest time between sets—is another form of increasing intensity. But again, you must use the method intelligently. Make your rest periods progressively (that word again) shorter as your competition or peaking date closes in. Naturally, you would need more rest between sets of squats than between sets of curls.

Number of Workouts per Week

Do not base your workouts on a seven-day schedule. Your workouts should be dictated by your body's needs instead. Of course, you also must take your job or school situation into account.

Because muscle that is worked hard takes considerable time to recover fully, you should not train the same body parts every day. Training frequency, as with schedules, sets, reps, and intensity, must vary with your training gains and recovery ability.

One of the most workable systems for adding muscle size is the every-other-day split. For the benefit of the uninitiated, the every-other-day split involves splitting your workout schedule roughly in half. Perform the first half on one day, and then rest completely (no training) the following day. The day after that, perform the second half of your routine. The following day, rest from all training. In other words, you do half your routine every other day, and you never work out two days in a row.

Choice of Equipment

There are not enough machines to take care of all of a bodybuilder's needs. You have to use free weights (barbells, dumbbells, and pulleys), if only for the necessary variety. And it just so happens that free weights are terrific for maximizing muscle mass.

Bear in mind, too, that most bodybuilders do use some machines to supplement their weight workouts. Variety is the spice of life —and the key to bodybuilding success.

Training Attitude

Not only must you have a positive attitude, but, unfortunately, your strategy for success must incorporate some selfishness. Relatives, friends, your spouse or partner must come to understand that workouts are something that you simply cannot miss. In addition, you must learn to see yourself as a bodybuilding success. (You just haven't collected your trophies yet.) You must maintain your positive attitude throughout every set of every workout. Concentrate on every exercise with all the dedication you can muster!

Number of Exercises

Your routine must be built from your own needs. You may want to do two or three exercises for a stubborn body part but only one "carry-along" exercise for an easy-growing area. At the least, the number of exercises you slot into your routine must be adequate to serve your basic growth needs. At most, they must stimulate growth without leading to staleness.

Amount of Weight

In a sense, the weight you use in an exercise is irrelevant, be it a 20-pound dumbbell or a 50-pounder. Some "cheat artists" can bounce-hoist huge poundages, yet they couldn't lift half the weight if they used strict exercise style.

One important difference between the bodybuilder who wins titles and the one who tries but fails is that the pro sees barbells as tools for gaining muscle, not as weights that have to be heaved up. Getting a weight up is not the aim. Making the muscles "feel" the weight is what we have to accomplish.

Order of Exercises

It is not a good idea to perform exercises randomly. Group body parts together so that one area is fully pumped before moving on to another area. Perform heavy (demanding) exercises early on in the routine, while you are full of energy. You can taper off a session by working the smaller muscle groups, like the abs or calves.

Milos Sarcev and Chris Cormier, in the spotlight to compare "lat spreads"

10
RECUPERATION

MENDING THE MUSCLE

A ny amount of training always carries with it a negative factor, in that it drains some of our resources. Therefore, adequate recuperation is essential for the bodybuilder. The faster we can recover from a strenuous workout and replenish our resources, the sooner we can train again. And the sooner we can train, the faster we will grow. On the other hand, continuous training without full recuperation will drive you to the worst sticking point ever. Your motivation will suffer, your muscles will shrink, brute power will ebb, and—hell of all hells—the pump will evade you.

Here's what my old friend, the late Vince Gironda, had to say about overtraining: "Overtonis is my expression for the condition caused by too many sets and too many different exercise combinations, for the overwork which causes muscle tissue loss, hormone depletion, weakness, a smoothed-out or stringy appearance, inability to produce a pumping effect, and general lassitude or weakness. Overtonis stops the central nervous system from pumping blood into capillaries that might otherwise rupture. It is a safety valve activated by hormone loss. Going past the pump (too many sets when the body is not used to it) is the most common cause of overtonis."

Overtraining first rears its ugly head the day after your workout. As you open your eyes the next morning, you feel decidedly groggy. In fact, you may feel disinclined even to get out of bed. You feel very tired. Your body is saying to you, "Boy, I really had a hard workout yesterday. I need more time to recover."

This tired, depleted feeling is not always accompanied by sore or aching muscles. What sore muscles usually imply is that the body is recuperating and the healing process is well underway. Of course, if there is any soreness present, the muscle has not fully recuperated, but at least the healing process is taking place. The general opinion is that a slight soreness in the muscles the next day is ideal. It's evidence that your workout routine got to your muscles. But if the soreness is extreme, then you have overdone it, and adequate recuperation will take longer.

Workout Wisdom

There is a fine line between stimulating your muscles with new exercises, more intensity, and added sets—

(Left) Tom Pattyn
(Above right) Flex Wheeler

all for the sake of keeping them growing—and the obvious mistake of going all-out with these tactics and reducing the likelihood of adequate recovery in a reasonable time period.

Let's take an example. Imagine that you have always done 6 sets of 15 reps with your calf work, and you decide that you need to jolt your lower legs into growth by trying something new. Well, you could add on 3 sets of donkey calf raises. That would give the calves a new exercise, and you also would be making the overall calf-work routine longer (more sets). But suppose you wanted to give your calves a real surprise. How about suddenly doing the new donkey-raise movement for 20 sets? Wouldn't that be better? Would such a change not shock them into growth? The answer, of course, is No! You would be lucky to be able to walk the next day—let alone boast additional lower-leg size—and you probably couldn't train your calves again for a week or more.

The high-intensity, low-sets workouts advocated by multi–Mr. Olympia Dorian Yates will definitely hasten recuperation. Once your body has become used to this method, your recovery from workouts will be quick. Yates says, "When I was using eight to ten sets for every exercise, I felt constantly tired. Progress was slow. Now, with my heavy-duty program, I recover very quickly after workouts."

Still under debate, of course, is whether full-intensity (heavy-duty) workouts are right for everyone. Flex Wheeler is one of the many champions who do not advocate training to failure. And Joseph Miller, a sports-medicine authority, had this to say in *Muscle & Fitness* magazine: "Physiological research indicates that the adrenal glands can become thoroughly oversaturated and exhausted if they are forced to overcome maximal stress for any period longer than 2–3 weeks. If high-intensity effort is sustained beyond this time, the adrenals will be forced into total remission, and training efforts at that point will illustrate this."

The aim, of course, is to train the body without fatiguing it beyond its power of swift recovery. If you do not "heal" between workouts, you will not grow. The possibility of overtraining looms ever present over the competitive bodybuilder (especially when doing leg work), so he must strive to perform the minimum amount of exercise that will stimulate continued muscle growth. Performing more sets than are necessary to induce growth will only delay the recuperative process. An abbreviated training routine certainly has its place in the bodybuilding world, if for no other reason than to effect better recuperation. It is only after full recovery has taken place that growth can occur.

If you follow a high-intensity routine (forced reps, negatives, and so forth), you will be fatiguing your body to a greater degree. This increased intensity causes your body to produce more lactic acid—hence, the greater fatigue. Therefore, should you adopt a more severe program, slide into it by steps. If you work out wisely by increasing the volume and intensity gradually, your recuperation likely will keep pace. There is no doubt that a body that is coaxed into handling more will also recuperate more quickly.

Detecting Overtraining

In *The Sports Medicine Book*, Dr. Gabe Mirkin lists the following indicators of overtraining:

Muscle Symptoms

Persistent soreness or stiffness in joints and tendons
Heavy-leggedness

Emotional Symptoms

Loss of interest in training
Nervousness
Depression
"I don't care" attitude
Inability to relax
Decreased academic work or performance

Warning Signs

Headache
Loss of appetite
Fatigue and sluggishness
Loss of weight and muscle size
Swollen lymph nodes in neck, groin, or armpit
Constipation or diarrhea
Sore throat

Mike Mentzer describes a simple method by which bodybuilders can check on their condition: "For decades one of the most popular methods used to detect overtraining was to monitor the morning pulse rate. Upon rising, the athlete would measure his pulse for 60 seconds. If it was 7 beats a minute faster than usual, a layoff or reduction in training was indicated."

Fatigue toxins also can be measured in order to determine whether you are in an overtrained state. One way is to have the enzyme levels in the blood measured, because damaged muscles release more of these proteins. Sports doctors who monitor their athletes carefully know that when the enzymes exceed a certain level, the athlete has to slow down.

However, an experienced bodybuilder is usually well aware of not having recovered from a previous workout. He feels it. Acquaint yourself with how you feel the day after your workout. Make notes in your training log. Stay in tune with your body.

The ultimate test of whether a bodybuilder is overtraining is that his progress has come to a halt. But before he concludes that the problem is one of overtraining, he had better be sure that he is working

Dorian Yates flexes with Flex Wheeler at the Mr. Olympia.

out on a progressive basis. If not, his no-gain status may be due to laziness or insufficient work.

Schedules

The number of times you should train each week depends in large part on your recuperative powers and your tolerance for strenuous exercise. Today, body science indicates that training each body part (for example, legs, chest, and shoulders) only once or twice per week is the fastest way to add lean muscle mass. This leaves us with several workable alternatives.

You can split your workout routine in half, and train the first half on day one, rest on day two, train the second half on day three, rest on day four, train the first half again on day five, rest on day six, train the second half on day seven . . . and so on.

Arnold Schwarzenegger, Ben Weider, and Franco Columbu

You can split your workout in two, and train the first part on Mondays and the second part on Tuesdays, rest on Wednesdays, and then train the first part again on Thursdays and the second part again on Fridays. This, unlike the previous routine, allows for a workout-free weekend.

An increasingly popular method of training these days is to work one body part per day for six days in a row. Naturally, this means that you are training your entire body only once a week. It does, however, work well for many, but you need a good variety of exercises to do full justice to each body part. Five or six exercises (4–6 sets each) are recommended.

Because we all have different physiological requirements as well as personal commitments, such as those having to do with jobs and families, we have to devise a workout schedule to accommodate our individual needs. Some choose to train two days in a row, followed by one day of rest. Another frequency pattern is to train three days in a row, followed by one day of rest, followed by two days' training and one day of rest. The possibilities are endless. The important point to remember is that each body part must be worked hard at least once a week. If you feel that your muscles are not responding to this frequency, then experiment with more frequent training. You can exercise each body part as much as three times a week.

Relaxation

What speeds recuperation? The first answer that comes to mind is relaxation. If we don't relax, full recuperation will be prolonged. Many top bodybuilders have learned the art of relaxation. They do it by turning their mind to other interests, such as art, music, conversation, yoga, poetry, religion, books, TV, or time spent with family and friends.

Many of us respond best to music. Maybe you do, too. So relax, put your feet up, absorb. Spend half an hour letting your favorite music wash over you. Inspiring and profound music will do the best job of relaxing you. But, whatever you select, don't be afraid of experiencing deep emotion. You want to be a whole man, not merely a muscle machine.

If the Mona Lisa rests your weary mind, get yourself a print. Maybe you like Picasso, Dali, or Van Gogh. Sculpture—why not? Maybe Rodin's *The Kiss* or Henry Moore's *Reclining Woman*? Reproductions are available. There is also the beauty of nature: fields, streams, mountains, and valleys. What could be more restful?

Do you like poetry? You can read it or listen to it on tapes as many bodybuilders do. Read some good authors: Thoreau, Wilde, Shaw, Russell, Emerson. Read about things other than muscles, and let your mind carry you to an entirely different space.

Some people believe in never running when they can walk, never standing when they can sit, and never sitting when they can lie down. Within limits, the overextended bodybuilder can take a hint from this philosophy. In other words, rest when you can. Enjoy a nap at odd times if at all possible.

Stress

Stress is the opposite of relaxation. When we are stressed out, our digestive system shuts down, our heart speeds up, our breathing deepens, and adrenaline and other hormones flood the body. Some stress is necessary and can be seen as a built-in survival mechanism preparing us to fight or flee.

Unnecessary stress, on the other hand, leads to wasted energy, and too much of it will keep you from becoming Mr. Olympia. You cannot expect to make regular gains in muscle size if you overwork, overplay, engage in frequent arguments, or generally burn the candle at both ends. Under this kind of stress, the body reacts poorly to training. Chronic stress drains the body of energy. In addition, the trainer loses his mental concentration and the ability to recover quickly from his workouts.

Sleep

One definite requirement for full recuperation is adequate sleep. As a hard-training aspiring bodybuilder, you need eight hours of sleep each night. It is possible to get by on less, but it is not ideal. Many bodybuilders keep fairly good hours on weekdays, only to party like crazy over the weekend. Scientists have concluded that we can get too little sleep one night and make up the balance the following night; in other words, we can borrow and pay back sleep, and suffer no adverse consequences. However, such practices are not advisable. Sure, you can party once in a while, but make it the exception rather than the rule. Generally, try to get those eight hours of sleep every night.

If you have difficulty sleeping because you are overtrained or anxious about an approaching contest, try some relaxing herbal tea to settle your nerves. You might even take a mild sleeping pill, but don't get into the habit of it. Don't exercise immediately before going to bed, for it will fire up your metabolism at a time when you want the opposite effect.

Sun and Fresh Air

Fresh air and sunshine can speed up recuperation. I am convinced that this is one reason why bodybuilders in California and Florida make such good progress. Sunshine in moderate doses also stimulates hormone production—a definite plus for the enthusiastic bodybuilder. But beware of heavy doses of sun—you don't want to get a sunburn.

Fresh air, breathed deeply, energizes the entire body. In fact, some consider an athlete's success dependent on the amount of oxygen he ingests. Didn't Arnold Schwarzenegger and Franco Columbu make the best gains of their lives when training at the original Gold's Gym at Venice Beach? They would inhale pure oxygen from divers' oxygen tanks between heavy sets of grueling barbell exercises. You may want to dismiss the benefits of fresh air for speeding up recuperation, but they are a fact of life. What's also certain is that stuffy, centrally heated surroundings will do the opposite.

Steroids

Sometimes I hear about the exhausting workout performed by some top-name bodybuilder and I recoil in disbelief. When I actually witness him training, though, I have to believe what he's doing there, right before my eyes. It should be said that many competitive bodybuilders can train in this manner because they aid their recuperation by taking anabolic steroids. *This practice is definitely not recommended.* Many serious bodybuilders have become very sick as a result of taking steroids without a proper doctor's prescription and the attendant monitoring of their vital functions.

Being a hard gainer myself, I have been sorely tempted, but I have never taken an artificial steroid, not one. Of course, I do have an insight into the matter that few in the sport can equal. As publisher of *MuscleMag International* and *Oxygen* magazines, I am the steady recipient of mail from readers who have had terrible experiences with these muscle-growth drugs. Here's one sample from bodybuilder Hank Zarco:

"I've been a bodybuilder for 31 years. In that time I have experienced all kinds of mishaps from not warming up, overtraining, improper eating, liquor, not enough sleep, and a lack of bodybuilding knowledge. But I still made great gains in strength and muscle quality.

"I was very proud of my strength. Weighing between 142 and 150, I could do 545 pounds in the parallel squat, 2 sets of 15 reps; wide chins, 8 sets of 8 reps, with 100 pounds around the waist; bent-over rows, 3 sets of 5 reps, 300 pounds; strict barbell curl, 160 pounds; one-arm French presses, 8 sets of 8 reps, 75 pounds.

"But I couldn't stand prosperity. I heard about steroids, and from that point on my health (internal organs) took a downward turn. I wasn't satisfied with a one-a-day prescription. When I gained a few pounds, I thought if one dianobol (5 mg) tablet could do that so soon, what would happen if I increased the dosage?

"For three years I took three 5 mg tablets a day. Of course, I won Mr. Illinois, Mr. Central USA, Mr. Mid-West and 2nd Mr. America in the short class (IFBB), Mr. Colorado, and a total of 67 trophies. But the price was high, because the steroid was beginning to take its effect.

"Every once in a while when I took a few weeks off from the drug, I shrank in muscle size till I went back on it again. Each time it was harder to get back into shape.

"Then my prostate began to suffer till I bled from the penis. Next it was my liver, my kidneys and my bladder.

"I was forced to stop steroids. Getting into shape was a chore. I could not get my muscles to respond. Heavy weights hurt me. It took longer to recover, and it seemed as though my tissues were not responding.

"My strength began to lessen each month, until I was using the poundages of a beginner. But I did not give up. I learned humility. I started using better form, more concentration, and the desire to find a better way for better results. This is when better things began to happen to me.

"I found a way for improving a set, exercise, and my own ratio for better recuperation. In short, I developed my own bodybuilding principles. I wish I would have known then what I know today, but that's life. Maybe other bodybuilders can learn from my experience (mistakes). I sure hope so."

Other Drugs and Diversions

During my many years in bodybuilding, I have known some champions and near champions who were smokers and drinkers. In their first year or two of training, their bad habits didn't appear to be detrimental to their progress. But then, low and behold, telltale signs began to show up. I would notice a gradual reluctance to perform heavy, high-rep squats and other demanding exercises. That was because both cigarettes and booze have a toxic effect, and it had worsened and caused nausea during the more vigorous movements. As time went on, these men's workouts became shorter and their recuperation took longer. Eventually, they could handle no more than an occasional workout, and sometimes only a few exercises. This is all because booze and nicotine had clogged up their system, tired them out, and robbed them of their vitality.

The so-called recreational drugs, like marijuana, are equally harmful, of course. Young bodybuilders can handle them for a while, but the ultimate burnout is as inevitable as death and taxes.

Some young bodybuilders are concerned about the possibility of negative effects from masturbation. In the not too distant past, masturbation was believed to cause everything from madness to blindness. Neither was true, of course, although excessive masturbation can make you overtired and adversely affect your training vitality and general recuperative ability.

Nutrition

Not to be overlooked is the role nutrition can play in maximizing your chance for complete recovery. At mini-mum, you should have the recommended daily allowance (RDA) of protein, vitamins, and minerals. If you feel that your workout recovery needs a boost, then I suggest you take a multivitamin (one-a-day-type) tablet. In addition, you also might benefit from daily doses of a high-quality protein powder, vitamins C, E, and B-complex, and some type of chelated mineral supplement.

Bear in mind, also, that if you are trying to gain muscle while losing fat, you may have cut your carbohydrate intake down too low. The days of precontest starvation are over.

At the very least, you need an adequate diet for fast and complete recuperation. Your recovery from demanding workouts is extremely important to your ultimate success. Neglect nothing that can help.

Aaron Baker and Flex Wheeler compare abs and thighs.

11

BODY-FAT PERCENTAGE

THE LEAN ADVANTAGE

They used to call it definition. Today, we use a different term: body-fat percentage. There is an undeniable movement toward ultimate definition, at least around show time. After a contest, most bodybuilders gain 10 pounds, others 30 or even 40. Usually, this does not manifest itself as rolls of fat, but as increased overall size and so-called thicker skin. When you reduce your fat percentage to less than 7 percent, your body takes on a whole new appearance. Not only do veins show up in minute detail, but cross-striations of the muscles become apparent. You will start to look like an anatomy chart—a picture of muscles with the skin stripped away.

This is a condition that has led to a good deal of controversy. Many find bodybuilders with an extremely low percentage of body fat to be repellent and unsightly. Nonetheless, most current followers of the sport actually prefer this "ripped" look, and it is the fellows with the least fat who are walking away with the top titles in bodybuilding.

The book *Physical Fitness of Champion Athletes,* by Thomas K. Cureton, Jr., computes the average fat percentages of various groups of athletes. Sixteen track-and-field athletes were found to have 11.88 percent of body fat. For 15 Danish gymnasts, the body-fat average was 10.36 percent, and for 21 Olympic swimmers, the average was 10.60 percent. All the groups tested had a median height of 69 inches (1.725 meters) at a weight of 160 pounds, and their average body-fat content was just under 11 percent.

It's been said that the average man has a body-fat percentage of 12.56, and that the minimum for men should be 10 percent, the minimum for women 20 percent.

John Grimek, at 195 pounds, had a body-fat ratio of 10 percent. Eugene Sandow had 11.26 percent. Among Olympic and power lifters, the percentages are somewhat higher. Doug Hepburn was recorded at 14.07, and Soviet mountain man Vasili Alexeev at 23.70.

With today's champions, unlike those of yesteryear, the body-fat percentage often fluctuates according to the stage of their training. It is not uncommon for a competitive bodybuilder to cut his body fat by two-thirds before a contest.

Years ago, an athlete or bodybuilder was praised and honored for possessing a body of harmonious,

(Left) The lineup
(Right) Clarence Bass

smooth muscular development. Today, the word "smooth" is used solely to deride a bodybuilder for being totally out of shape. Times change.

Incidentally, some of the ancient Grecian statues that presumably represent the Greek ideal of male perfection also have been scrutinized with regard to their body fat. The famous Farnese Hercules is estimated to have 11.96 percent, the Apollo Belvedere 11.76, and Myron's Discobolus 12.06. If these statues were real men, they probably would not do well in the Olympia contest, where the top six men have averaged a body-fat count of less than 5 percent.

Bodybuilder-turned-weightlifter Clarence Bass got his body fat down to 2.4 percent, and even between contests holds his body fat down to less than 6 percent. Bass has written two excellent books, *Ripped* and *Ripped 2*, which both report in detail how he achieved ultimate muscularity without undue hunger or pain. Although he rarely indulges in pizza, ice cream, or other high-calorie junk foods, he does eat cereals, raisins, eggs, and potatoes. Like Frank Zane before him, and Vince Gironda before Zane, Bass has set a trend. Each defined this new ideal.

Why has this current trend of being almost totally fat-free caught on the way it has? One reason is that it enables us to see muscles that we never knew we had. Few physique champions of the distant past could show shapely, delineated serratus muscles and incredibly separated thigh muscles. Today, to compete, all your muscles have to be diamond-sharp. That includes the muscles of the upper thigh and lower back, two areas from which it is difficult to eliminate all fat. Maybe trends once again will return to the 11 percent ideal, but, for the moment, low fat and cross-striations are the order of the day. This even includes the glute muscles, or buns. To win a contest today, those glutes must be striated.

Diet

How do you bring your body fat down to a very low percentage? The simple answer is you eat less, but it's really more complex than that. Basically, you must cut calories progressively and reduce simple sugar and animal fat in your diet.

It is advisable for the food you eat to be as near to its natural state as possible. Foods with natural fiber keep you leaner and fitter than dense-calorie foods. Eat whole-grain breads, cereals, fruit, vegetables, fish, cheese, organic and white meats, and skim milk. Stay away from the processed, chemically treated, artificially flavored, brilliantly colored garbage often found at the local supermarket.

Thyroid

Many physique men who do not like to diet rigidly resort to taking a substance called thyroid, which has the effect of speeding up the metabolism. As you may expect, the body's own production of thyroxin often will stop when large doses of thyroid are supplied artificially, even under medical supervision. In many cases, this becomes irreversible, so that the bodybuilder has to take thyroxin artificially for the rest of his life. This will not serve to make you a happy camper.

Diuretics

Another practice is the taking of diuretics in large amounts. LaSix is one such product, and it causes the body to eliminate a great deal of fluid in a short time. With LaSix, some vital minerals also may be flushed out of the system, like potassium, which is needed to regulate the heartbeat, among other things. Bodybuilders have even been known to die from the indiscriminate use of diuretics. One star bodybuilder told me in private that he took several tablets of LaSix a few days before a Mr. Olympia event. On stage, his body started to burn up. He got cramps in his legs, and it became hard for him to breathe. Ultimately, he gulped down water backstage to keep himself from passing out.

Although steroids tend to have a relatively slow effect on the body, eventually contributing to heart problems, prostate malfunctioning, and liver disease, it is now understood that the casual use of prescription diuretics before a contest can cause unconsciousness and even death.

Vitamins

Choline and inositol, members of the vitamin B family, are known as aids in the redistribution of fat. At least, they are safe and have no known toxic effects, unlike the products mentioned before.

You can buy choline and inositol, separately or together, at your local health food store. They were publicized first around 1960, when it was reported that most of the Muscle Beach stars took these vitamins prior to a contest or posing exhibition. Further ex-

Rory Leidelmeyer

Aerobic exercise stimulates the production of enzymes that convert fat to energy. The more fat-burning enzymes you have, the better you can use up, or burn, excess flab. Not only does aerobic activity burn calories better than anything else, but it also increases the body's capacity for burning fat. A long-distance runner is a veritable fat-burning machine. The reason this is so is that the aerobic activity keeps the heart-pulse rate below 80 percent of his maximum. To estimate your maximum heart rate, subtract your age from 220.

Unlike weight training, which can boost your heart rate temporarily to near maximum, walking keeps your heart rate well under 80 percent of your maximum. According to Bass, in *Ripped 2*, steady, prolonged walking comes as close as anything to being a pure fat-burning activity.

Needless to say, you should limit your aerobic exercise if you are trying to gain weight. Even during regular maintenance training, you should not overdo this form of exercise, because it can detract from your bodybuilding gains. Some degree of common sense has to be used so that you balance your muscular development with your aerobic fitness. If you are fit and well muscled, you really have a double advantage: you have extra fat-burning enzymes to help you stay lean and extra muscle mass that in itself increases the metabolic rate.

perimentation led to the conclusion that choline and inositol appeared to work for the physique contestant if their intake was combined with a strict diet. Certainly, no one could eat like a pig and keep his weight under control simply by taking these substances.

Today, there are numerous "fat burners" on the market that bodybuilders swear by. They are produced by the big supplement companies, such as TwinLab, EAS, Pro Lab, and MuscleTech. Even *MuscleMag* has its own Formula One line that includes a very workable "fat burner."

Aerobic Exercise

Weight training may be unparalleled in its potential for building up the skeletal muscles of the body, but it is not particularly good for burning fat.

"Aerobic exercise should be included in the program of any bodybuilder interested in staying lean," says Clarence Bass, who, probably more than anyone else in competitive bodybuilding, has made a science of staying lean while holding muscle size. Stationary-bike riding, road cycling, slow jogging, swimming, and walking are all aerobic exercises that burn calories through prolonged, low-intensity effort. Weight training is not aerobic but anaerobic exercise, and, unlike aerobic exercise, it does not result in a steady need for oxygen and a considerably stepped-up heart rate.

Joe Spinello of Montreal

12

METABOLISM TRAINING

CREATING THE ANABOLIC STATE

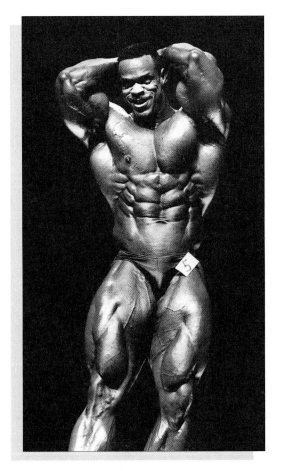

W hat is the metabolism? What is your metabolic rate? Your metabolism consists of all the chemical processes by which your body produces energy and assimilates new material to maintain, replace, and build up its cells. Your metabolic rate is the speed at which your body burns up fuel. Just like a car, your body has a tick-over speed. When it is running fast, it will burn up a great deal of fuel. At a more moderate pace, it uses less fuel. Like a car that continues to idle when it is not moving so long as the engine is running, your metabolism still is working when you sleep.

People have many misconceptions about the metabolism and its relationship to bodybuilding. What is important to understand is that the successful bodybuilder does not try to speed up his metabolism to a super-accelerated rate (unless he wants to shed fat very quickly), nor does he try to slow his metabolism to a subnormal level.

In nature, we can observe two creatures with a high metabolism that, due to their genetic makeup, are always in need of food. The shrew is in endless pursuit of food. When you look at this minute creature, you can see its system pumping away and shaking its tiny body like a battery-driven toy. The other insatiable creature is one of nature's marvels, the tiny hummingbird. Its super-fast wing beats enable the hummingbird to hover in midair while stealing its vital nourishment from plants. From a study of such high-metabolism creatures, however, we can conclude that a fast metabolism is of little use to the bodybuilder.

So, what about a slow metabolism? What creatures can nature offer us in that direction? The koala bear, the elephant, the rhino, the sloth. A slow metabolism may be conducive to gaining weight, but it in no way effects a rapid increase in muscular tissue, which is the ultimate aim of the bodybuilder.

There are two aspects to the metabolic process: the anabolic (or building up) and the catabolic (or breaking down). Both are constantly taking place in your body. The most desirable state, and one that you can train for, is a positive ratio in favor of the anabolic (building-up) aspect.

(Left) Paul Dillett displays amazing abs and "intercostals."
(Right) A chest pose from Australia's Lee Priest

The Super-Fast Metabolism

Many young people are plagued not only by skinniness but by a super-high metabolism. Whatever they do, they just cannot gain weight. Even when they drink huge quantities of milk and eat tons of nutritious food, they manage to burn it up. Their weight just does not increase. Then finally, one fateful (or glorious) day, their metabolism seems to normalize, and suddenly their bodybuilding efforts and generous food intake begin to show results.

Our metabolism slows down as we age. Then the fat starts to scttle around our bones, and we yearn for those earlier days when we had an accelerating metabolism.

The only other way to slow down your metabolism, besides growing older, is to purposely practice relaxation—real relaxation. Try to check on yourself during the day. How do you watch TV? Do you lean forward in your chair, or do you sit back comfortably with your feet up on a padded stool? Get the idea? Relaxing after a meal is particularly important, even if for 10 minutes. Try to rid yourself of tension and stress, mental as well as physical.

The Super-Slow Metabolism

If you feel you have an especially slow metabolism— if you tend to be overweight and lethargic—you can take steps to stimulate your metabolic processes so as to normalize it. When that begins to happen, your digestive processes will accelerate, your glands will secrete more, and your hormones will be stirred up.

How can you speed up your metabolism? By fitting exercises that stimulate the metabolism into your schedule. For a while, you will have to put abdominal training, calf work, and arm exercises aside. The true stimulators of our metabolic functions are the movements that work the bigger muscle groups— in other words, squats. It's recommended to squat with 20 repetitions, and the last 8 repetitions should be forced reps. Make a point of breathing deeply between reps.

No Recent Progress?

When a bodybuilder has been unable to make any progress at all—especially in his late-beginning or early-intermediate stages—his gains will accelerate enormously with a heavy, high-repetition squat

program just two days a week. He does not have to use many other exercises. Numerous s.uccessful cases have included only the wide-grip chin and bench press along with the squat training for their metabolism.

Virtually all the top bodybuilding champions have used heavy, high-rep squats in their training to stimulate their metabolism and give their body overall size. Once this size has been reached, many champs find they don't need to use this form of training anymore. They can get by on hack slides and leg extensions, as well as other less strenuous leg movements. Ask Paul Dillett, Nasser El Sonbaty, Lee Priest, or Michael Francois. Even when training for a Mr. Olympia title, they may need only four to six weeks of strenuous squatting to bring their thighs and their metabolism back up to peak condition.

Remember that this metabolism training is not necessary as long as your current bodybuilding routine is developing your muscles at a satisfactory pace. This mode of training is designed specifically for

Nasser El Sonbaty flexes his pecs.

Michael Francois

the hard gainer who needs to boost his anabolic state. Virtually all hard gainers who have applied this principle seriously over a period of four to nine weeks made significant progress. What is more important, they made this progress when every other form of training had failed.

It should be added here, however, that no one can go beyond the limits of his genetic endowment. At the same time, what's fairly certain is that no Mr. America, Mr. Universe, or Mr. Olympia yet has completely reached the muscle-size limits set by his own genetics.

Is it time for you to start squatting?

69

13

ULTIMATE NUTRITION

MUSCLE BUILDING AND THE FOOD FACTOR

Together with progressive resistance exercise and adequate rest, nutrition is an essential requirement for bodybuilding success. Today, this is so more than ever, because it is the bodybuilders with the least amount of fat and the highest definition who are winning the contests. There is the accepted notion now that "we are what we eat," so almost everyone has something to say about food and diet. The ideas they propound run the entire gamut, from the acceptance of junk food to the necessity of prolonged fasting. Needless to say, neither extreme is recommended.

Traditionally, bodybuilders have gone in for ingesting huge amounts of protein—steaks, cheese, poultry, fish, eggs, nuts, and, of course, protein supplements. But with the publication of Dr. Nathan J. Smith's tome *Food for Sport*, which proclaimed, "It is important for athletes to recognize that their athletic activity, although it may require a high energy expenditure, will not significantly increase their need for protein," the protein-overload theory has lost some of its popularity. Still, let us not overreact. Even the most enthusiastic member of the low-protein brigade has to admit that the would-be champion bodybuilder does indeed need a little more protein than the layman, or even the competitive athlete.

Yet, protein is the least efficient source of energy, and too much of it may result in a slower rate of recovery and give you superfluous calories that turn into fat. It is far better to be on a high-carbohydrate diet.

Carbohydrate is the main fuel for muscles. It becomes glucose in the blood and is stored by your muscles and organs in the form of glycogen. During and after concentrated workouts, your glycogen becomes depleted. Because your recovery is largely dependent on the restoration of glycogen in the muscles, a diet high in carbohydrate speeds the restoration process.

According to Clarence Bass in his book *Ripped*, "Studies have shown that a high carbohydrate diet fully restored the glycogen in the muscles after 48 hours, while a high protein and fat diet left glycogen levels below par even after five full days. Clearly, a diet high in carbohydrate does the job of getting you ready for your next workout."

**(Above right) Paul Dillett
(Left) Apples, with all their fiber, are a lot better for you than store-bought apple juice.**

71

How much protein do we need for our muscles to grow? Actually, muscle building is a relatively slow process, so we don't need all that much. For a 154-pound man, the National Research Council sets its recommended daily allowance of protein at 70 grams (1 gram per 2.2 pounds of body weight). This is a very generous allowance, as some studies have shown that people can stay very healthy on a considerably smaller intake of protein. But if you take more, then do it sparingly. This will not only help to keep the fat off, but it will also save you a bundle of money.

Lest you think I am particularly down on protein, let me be the first to say that you don't need much more of any extra food—protein, fat, or carbohydrate—when you are trying to gain muscle. What is important is that you eat frequently. Five or six small meals are infinitely superior to three large, gut-bustin' gourmet extravaganzas. Small meals maintain your blood-sugar level, keep you from getting hungry, and prevent the discomfort of digesting large meals. It is an antiquated idea to stuff yourself to fill out. The result invariably will be added fat, not muscle.

Of course, there are people who must eat a lot. For example, if a budding weight trainer has a high metabolism (the rate at which his body burns fuel while resting), his food intake must be tailored accordingly. If you burn up 3,000 calories a day, then something more than those 3,000 will be needed for you to gain weight. There are individuals who require 5,000 calories a day. If they eat more than that, they will gain weight. If they eat less than that, they will lose weight. The art is to find out by trial and error how many calories you need to gain or lose weight slowly. Doing either too quickly will give you less satisfactory results.

It is generally accepted that a person who leads a moderately active life has a daily need of about 15 calories per pound of body weight. Your food intake should maximize your chances of bodybuilding success. Misuse your nutritional regimen, and you will either fail to gain muscle mass or else cover what muscle mass you do develop with an unattractive layer of fat.

A balanced diet (how many times have you heard this phrase?) is the best way to gain lean muscle mass, or muscle without the encumbrance of fat. It is true that a balanced diet can mean different things to different people, but in essence it should include nutrition from each of the following food groups (low-fat items are best):

- Milk (milk, yogurt, cottage cheese, cheese)
- Meat (beef, veal, lamb, pork, fish, poultry, eggs, etc.)
- Vegetables (fruits, vegetables, legumes, nuts)
- Grains (bread, cereals)
- Fats (butter, vegetable oils)

About 65 percent of your diet should be made up of grains, fruits, and vegetables; the remaining 35 percent should come from the milk and meat groups.

Calories are important, of course, but do not simply choose a program of low-calorie foods when you want to lose weight, or of high-calorie foods when you want to gain weight. Foods should never be evaluated solely by their calories. Dr. Jean Mayer of the Harvard School of Nutrition says: "A proper diet must provide all necessary nutrients in sufficient amounts, be palatable, easily available from the viewpoints of economics and convenience, and be balanced in calories to produce the desired caloric deficit for weight loss or additional storage for weight gain."

You must aim for the best possible return for every calorie you consume. This is why you must avoid "calorie-dense foods," those unforgivable concoctions prepared or manufactured to appease the whims of taste, with little or no regard for sensible nourishment. Calorie-dense foods, also known as junk food, are characterized by their obscene preponderance of chemical additives, coloring, and preservatives, not to mention their lunatic levels of salt and sugar.

As the name suggests, calorie-dense foods provide calories in abundance, but they do not really satisfy the appetite. As a result, you tend to overeat, for the more of this garbage that you eat, the more you want. And its nutritional value is awesomely deficient.

Sugar and butter are perfect examples of calorie-dense foods, and, one way or another, they appear in unnecessary quantities in just about every food man has attempted to "improve." Other calorie-dense foods (though to my mind they hardly merit the name food) are gravies, cream, jam, canned fruit, pastry, cookies, shortening, candy, chocolate, jelly, soft drinks, processed cheese, potato chips, regular breakfast cereals, salad dressing, canned soup, ketchup, ice cream, and crackers.

It is always amazing to me how a natural, wholesome product such as bread can be devitalized by reducing its natural vitamin content, removing its bulk fiber, bleaching it, dosing it with preservatives so that it stays squeezy-fresh on the shelf, and adding vitamin

D or whatever, and it is then claimed to be "vitamin-enriched."

Most packaged or canned foods are processed highly and move through your system poorly, are totally unbalanced nutritionally, and make you want to overeat. How the government can allow the big food companies of the world to continue putting out these products is beyond me. Until the authorities have the sense to legislate this nutritional hocus-pocus out of our lives, you will have to watch out for yourself and keep away from all calorie-dense foods.

On the other hand, high-fiber foods are almost always low in calories, because fiber contains almost none. High-fiber foods are the "miracle foods" that man couldn't invent. They control calorie consumption and actually reduce the number of calories you absorb from the other foods that you eat. The more you keep to a high-fiber diet, the leaner you are likely to be, and with a lower percentage of body fat your muscle mass will be more impressive.

Now that you understand the value of fiber and the worthlessness of junk food (which is loaded with fat, sugar, and zillions of unpronounceable additives), try to make a point of eating correctly. The best sources of dietary fiber are whole grains, fruits, and vegetables. What you must realize is that when I say whole grains, I mean whole grains. A cereal label may claim that a product is made from pure, wholesome wheat or corn, but this does not mean that it is unadulterated. More than 90 percent of the packaged breakfast cereals are simply overpriced, sugar-loaded junk. Read the labels. In the same vein, do not think that canned fruits are acceptable. They are not. Look at the sugar content. And what about those other additives?

Avoid drinking fruit juices if you want to keep your body-fat content low. They pass straight into the bloodstream and are stored readily as fat—and you lose out on the fiber. It's far better to eat an apple in the way nature intended, with all its fiber, than to down a glass of crystal-clear store-bought apple juice.

Bread and potatoes are singled out invariably as fattening foods, and, served the way most people eat them, they are. Most bread is devitalized, and then we have the habit of covering it with calorie-dense butter and sugar-loaded jellies or jams. On the other hand, whole-grain bread made with natural ingredients is a particularly useful food for the bodybuilder and should be eaten in moderation every day. But even whole-grain breads may have added sugar and salt, so look at the

Lee Apperson and Debbie Kruck

labels to avoid these ingredients. It's the same situation with potatoes. We submerge them in fat and make them into French fries, or serve them with butter or sour cream. A baked potato, as opposed to the greasy French fry, is very nutritious. But please hold off on the butter and sour cream.

Your fat intake—and make no mistake about it, some fat is needed in the diet—can come from other sources, such as milk. You should drink some of it (preferably skimmed) every day. Milk is a wonderful nutrient. Those people who cannot digest it because of lactose intolerance can add the enzyme lactose to their milk. A product called Lact-Aid is made for this purpose and can be bought at drugstores and health food stores.

Milk is high in calcium, a very important mineral for the bodybuilder. Not only does it keep your bones strong, but it also helps your heart to pump and your brain to think, and it is vital for muscle contractability. If your body is not furnished with adequate calcium (three glasses of milk a day is recommended), it will rob your bones or your teeth for what it needs. A regular deficiency can cause osteoporosis (weakening of the bones), a disease suffered by 15 million North Americans.

Foods other than milk that are high in calcium include yogurt, cheddar cheese, green leafy vegetables (collards, broccoli, mustard greens, cabbage, kale, and spinach), dried beans, almonds, and Brazil nuts.

An enemy of the bodybuilder, especially when he is trying to cut up before a contest, is sodium. Yes, the dreaded salt. Certainly, too much of it predisposes one to high blood pressure. I also recall Dr. Albert Schweitzer stating that he felt an excessive salt intake was one of the principal causes of cancer. But, in addition to this, the bodybuilder needs to know that one part of sodium holds 180 parts of water. This may explain how some bodybuilders, in spite of rigorously reducing their calories, can come to a contest looking bloated.

A good rule of thumb for the bodybuilder, and for the health enthusiast in general, is to use table salt very sparingly. All people require salt for normal health, but our daily requirement is extremely low. In fact, deficiencies are rare because most of our foods contain sodium. You probably will get enough from ordinary, untreated fruits and vegetables.

It is quite safe to say that many people consume more than 300 times more salt than they need. This is partly because of its function as a preservative. As with sugar, salt helps to prevent food from going bad. Thus, sodium is pumped into just about every edible thing available. Check the labels, and there it is: sodium. A McDonald's Big Mac contains 1,510 milligrams of sodium, and a Kentucky Fried Chicken has 2,128 milligrams. When you consider that the average person often adds more salt to such foods with ketchup or other sodium-loaded sauces, you can assume that millions of people eat between 5,000 and 15,000 milligrams of sodium a day. Even cottage cheese, a longtime "food of the bodybuilder," contains 850 milligrams of sodium in just one cup.

Many bodybuilders tend to feel that more is always better, and they apply this not only to their muscle size but to the amount of food they eat. The problem is, of course, that sooner or later more food leads to more fat, and eventually you will be faced with the arduous task of taking it off. There is absolutely no evidence to support the theory that gaining fat hastens muscle growth. Stuffing yourself with every food in sight will help you gain weight, but there is no advantage whatsoever in providing your muscles with more nourishment than they require.

In bodybuilding, you must be realistic. A gain of one pound of lean muscle each month is commendable, certainly after the first year of bodybuilding, but it is extremely rare. If you train hard enough (and that is the first requirement), then it is possible, by applying a few nutritional facts, to calculate roughly how many calories you will need to achieve a monthly gain of one pound of lean muscle.

Simply put, a pound of muscle contains 600 calories, a pound of fat 3,500. There are, obviously, many more calories in a pound of fat, and fat development requires no outward stimulation, such as the progressive exercise needed to generate muscle. The reason why fat contains so many more calories than muscle is that the water content is only 15 percent, whereas muscles contain some 70 percent. There is also a wide difference in lipids (cell components high in calories) between the two. Muscle contains only 6 percent lipids, whereas fat contains 70 percent.

To stimulate one pound of muscle growth each month and make a total gain of 12 pounds in a year, you would have to increase your caloric intake by 600 (the number of calories in a pound of muscle) multiplied by 12 (the number of months in a year), or by 7,200 calories a year over and above the amount needed for current weight maintenance. This is 7,200 calories in one year, not one day. To ascertain how many additional calories you would require daily to gain pure muscle at this rate, simply divide 7,200 by 365 (the number of days in a year) and you will come up with approximately 19 extra calories a day. That's all you need to develop additional lean mass at the rate of one pound a month. It may be worthwhile to exceed this amount for insurance, but not by a large margin. Otherwise, you may become fat.

For all of these reasons, I do not agree with the concept of bulking up. By definition, bulking means the addition of size and weight at any cost, even though most of it will be fat. This was advocated in the old days of bodybuilding.

When I first became interested in the sport, my introduction to its nutritional side came from a Charles Atlas course. I was a mere kid then, living in Britain. The course recommended drinking several quarts of milk each day, plus eating a generous portion of eggs, fresh citrus fruits, and daily steaks. At the time, this advice could have come from Mars or Jupiter. Postwar rationing was still in effect, and most Britishers were limited to 6 ounces of meat per week. Eggs were a rarity, and I, for one, had never seen a grapefruit, or-

The back display is essential.

ange, or lemon. When rationing was removed, every bodybuilder in Britain went crazy. Milk was the main bulking agent. Many a bodybuilder put on 20 or 30 pounds drinking milk. Only later did we realize that more than half of what we had gained was fat. No matter, we refused to believe it and simply called it bulk.

I know that some men prefer the bulked-up look. They do not like definition. They just want to be big. If this is your idea of perfection, then simply drink a quart of milk 2 hours after each meal. Train two times a week for an hour and a half on six basic exercises: the press-behind-neck, squat, barbell rows, bench press, curls, and parallel-bar dips. Presto, you will get bulk—and you're welcome to it.

Nutrition can be as simple or as complicated as you want to make it. To recap, when a bodybuilder wants to gain weight, he must make sure he selects foods from the various food categories and eats sufficiently to nourish his muscles. Of course, he must also subject his body to vigorous progressive resistance every 48 hours. To lose fat, you must continue to exercise so as to maintain the status quo of muscle development, while progressively restricting your overall caloric intake. In addition, you have to watch your sodium.

Don Long

Not Recommended!

Here's a typical daily menu of the average North American:

Breakfast

	Calories
1 large glass of orange juice, or bowl of cornflakes	120
3 slices bacon	78
2 eggs fried in butter	202
2 slices white bread with butter and jam	275
Meal total	675

Lunch

Big Mac	561
French fries	214
12 oz. cola	145
Meal total	920

Dinner

4 oz. T-bone steak	535
Baked potato with pat of butter	240
° cup peas	56
Side salad, 1 tbsp. thick dressing	165
Apple pie and ice cream	395
Meal total	1,391
Meal total for the day	**2,986**

Snacks, gum, beer, candies, and so forth can easily add another 600 to 800 calories.

Recommended

Now here is a typical natural fiber menu for a day:

Early Morning Snack

	Calories
° grapefruit	45

Breakfast

1 cup oatmeal or rolled oats	130
1 tbsp. raisins	80
2 tbsp. bran	33
1° cups raw whole milk	225
1 sliced apple	100
Meal total	613

Lunch

Two-egg whole-wheat sandwich (no butter)	224
Mixed salad, 1 bowl (squeezed lemon to add taste)	150
1 orange	65
1 cup plain yogurt	150
Meal total	589

Midafternoon Snack

1 oz. cheddar cheese (1-in. cube)	116
2 oz. unsalted cashews or peanuts	200

Dinner

1 whole-breast broiled chicken	310
1 baked potato (no butter or sour cream)	90
1 cup green beans	30
2 whole steamed carrots (medium size)	40
1 cup fruit salad (no sugar, any combination of strawberries, tangerines, peaches, banana, blueberries, raspberries, grapefruit, apples, and pineapple)	80
Meal total	866

Evening Snack

1 bran muffin (low-fat)	85
1° cups skim milk	150
Meal total	235
Meal total for the day	**2,303**

Also Recommended

Here is another ideal menu for a day:

Early Morning Snack

1 banana	120

Breakfast

2 boiled eggs	160
2 tbsp. raisins	160
2 slices whole-wheat bread	144
1 cup mixed fruit salad	80
1° cups milk	225
Meal total	889

Lunch

3 oz. cold, sliced, lean roast beef	250
1 oz. cheddar cheese (1-in. cube)	116
1 medium tomato	35
2 carrot sticks	20
1 celery stick	5
1 slice whole-wheat bread	72
1 large apple	120
2 glasses water	0
Meal total	618

Midafternoon Snack

1 carrot-raisin muffin	85
1° cups milk	225

Dinner

Fresh vegetable soup (1 bowl)	100
Salmon steak (6 oz.)	300
Brown rice (° cup)	100
Broccoli (2 stalks)	90
Fresh corn	70
2 broiled tomatoes	70
1 cup plain yogurt with tangerine slices	190
Meal total	1,230

Evening Snack

1 banana	120
Meal total for the day	**2,857**

A wholesome, natural diet is far more filling than the average diet, and infinitely superior nutritionally. Forget processed, chemically treated foods. Keep it natural and fresh. If you have an inclination to be overweight, or if you are a bodybuilder who has already achieved an acceptable degree of muscle mass, you may find that your diet has to be "cleaner" than the previous suggested menus. "Clean" foods, as bodybuilders call them, include grilled poultry, oatmeal, whole-grain (no additive) bread, egg whites, baked potatoes, rice, water-packed tuna, steamed or raw broccoli, celery, lettuce, cauliflower, cabbage, apples, tomatoes, and so forth. Typically, an intermediate or advanced bodybuilder will eat five or more meals a day of something like chicken and rice or baked potato and tuna. It can be boring, but it works.

Terry Mitsos

14

DERAILING THE STICKING POINT

REGULATING THE PHYSIOLOGICAL PROCESSES

What happens when, after a layoff, you jump right back into hard-core bodybuilding? It happens to every guy without exception! You gain muscle size rapidly. But after growing like crazy for a while, your growth rate slows down and finally stops altogether. You continue to go all-out, eat well, and rest sufficiently, but the result is always the same—you're stuck! Nothing happens. Your muscles simply do not grow bigger. It is no exaggeration to say that 95 percent of all active bodybuilders are currently at such a "sticking point" in their training.

Progressive resistance is a stress, and when you first subject your body to this form of stress, it reacts quickly, preparing itself for further stress by strengthening and enlarging its muscles and tendons. But the body will do only so much. It will not continue to add muscle size unless there is a very good reason. Merely repeating what you first did to stress or shock the body is not enough to keep your body growing and growing. You must add new stresses, or increased-intensity shocks, for there to be progress.

Before I go any further, I should point out that you should expect to have periods of slow or nonexistent growth. At such times, the muscles consolidate their gains, or may even lose a fraction. During these periods, however, the body is building a base from which you can leap to the next level of muscular achievement.

There is nothing wrong with sticking points, provided they don't happen too frequently and they don't last too long. Nonetheless, overly frequent and overly long sticking points can be avoided.

Rest

If you do not allow your muscles sufficient time to recover after a workout, you will drive yourself to a sticking point that will be difficult to overcome. Someone who is always on the go will never build super-large muscles. If you are the type who plays tennis before your workout and goes dancing afterward, then wave good-bye to those 20-inch guns you have always wanted. They won't come this year, or likely ever.

I'm not saying that you shouldn't engage in other sports, but when the time comes that you want to maxi-

(Left) Claude Groulx of Montreal
(Above right) Hamdullah Aykutlu

mize your gains, then your extra activities must go. Later, when you have achieved your goal, you can bring them back into your schedule.

Even today with split and double-split routines, most bodybuilders find it most beneficial to train only every other day, leaving a complete day of rest after each workout.

Barbell and dumbbell training is the most severe form of exercise we have invented to punish our muscles. If you bench-press 200 pounds 10 times, in a mere matter of seconds you will have lifted some 2,000 pounds! Believe me, after a complete weight-training workout in which you likely will lift hundreds of tons by means of your repeated sets and repetitions, you will need . . . rest.

Pavol Jablonicky

"Shock Treatment"

One well-known IFBB pro bodybuilder was having trouble building his back. In the relaxed position, it just didn't compare favorably with the backs of the other pros. He didn't know what to do. Heavy rowing hurt his lower back, and he was already performing loads of chins. After talking the problem over with his trainer, our pro was given not one back routine but three! They were to be followed in succession, a different one being used for each back workout. "It will keep your muscles guessing," said his trainer. In *MuscleMag International*, Reg Park writes that in order to get his calves to respond, he had to jolt them with a different routine every day. Stubborn muscles must be surprised frequently to nudge them out of complacency and get them to grow bigger and bigger. Arnold Schwarzenegger has said, "I change around my exercises from time to time, and even perform them in a different manner . . . to shock my muscles into growing."

Sameness leads to boredom. Variety will bring about reaction. Of course, you could try to shock the muscles by never performing the same routine twice. Steve Reeves reportedly followed this principle to some degree. Nonetheless, the general consensus is that your schedule should be repetitive to some extent in order to stimulate regular exercise progression. In some deviously planned manner, what you need to do is overwhelm your muscles with a complete change of pace, a new exercise, considerably more or fewer repetitions,

a change in your exercise sequence or in the frequency of your workouts, or some other new mode of training. The prime requisite is that the change be sufficient to make your muscles react.

Layoffs

Like anything else, layoffs are beneficial if used properly. Taking layoffs too frequently will probably doom you to failure, but there are good reasons why you should take a break from time to time. A few days' rest—a week if need be—gives your body time to accumulate nutritional and nervous energy. The point is this: the body must be in good condition to benefit from exercise. Your body's reaction to progressive weight training is what it's all about. And your reaction to heavy exercise may be nil after a couple of months of training if you never give yourself a break. Timely layoffs, on the other hand, can keep your muscles growing.

Workout Duration

Gradually increasing the length of your workout (without increasing the length of the rests between exercises) can serve to increase the overload on your muscles, and thereby make them grow bigger. However, there's a point at which workout duration becomes too long and results start to regress. At that point, you must adopt a new, shorter program. Then, using your new routine, you increase its duration gradually to bring about sustained muscle growth.

The Carry-Along Principle

Perhaps this idea is best known as specialization, a technique that bodybuilders have been using successfully for many years. It often has been said that you cannot do justice to all your muscle groups all the time. That is why many men pay far more attention to one particular muscle group than to the others. They are concentrating on one area in order to bring that area up to par.

Say, for example, you had weak deltoids. It would make sense to start your workouts with heavy press-behind-neck, seated dumbbell presses, lateral raises, upright rows, bent-over laterals, front raises, and so forth. But you couldn't reasonably expect to use as many exercises for all your other body parts, so you would merely "carry them along" with one basic movement

Mark Erpelding, Gold's Gym, Venice

that prevents the muscles from actually losing size and tone. This way, you don't push your entire body into a state of overwork, but you just push one particular area to a new plateau. When that has been achieved, you may want to resume a more balanced schedule or concentrate on some other part of your body.

Holding Back

Cycling—or holding back for the sake of progression—is another method of derailing a sticking point. In a sense, this process of building up gradually to peak performance is an alternative to a layoff. Even so, a complete rest from all training is a good idea once in a while.

At the beginning of a cycle, you should deliberately "hold back" in your efforts, so that you can later make a steady progression. Instead of curling a 120-pound barbell all-out for 5 sets of 10 reps, hold back on a couple of sets, or stop when you know that you could do a further one or two reps. There is no need to perform more exercises or tougher ones than are necessary to maintain an upward growth pattern. Ultimately, as you close in on your peaking period, you will be pushing all the buttons for utmost intensity. But while you are at the beginning or middle of a training cycle, you should be consciously controlling how much energy you are putting out. Take pride in this control. Make your workouts quietly optimistic. Chaotic and panic-driven workouts can lead only to staleness.

Going for a New Plateau

One thing is for sure: you cannot aim for and reach a new level in muscular development without first imagining it. Your mind is the key to all significant progress. With iron-clad determination, not only can you overcome a sticking point, but you can make new and seemingly impossible gains.

You first have to make sure that your body is in the proper condition to manifest a substantial gain. You must be tuned up, yet not overtrained. Your body must be used to heavy, tough training, but not to the point of near exhaustion that puts you in a negative-growth phase. The base from which you lift off to a new plateau must be as solid as a rock. Your nervous system must not be shaky. Your food intake, supplementation, sleep, and rest must all be set for a new training push. You have to be fully prepared if you are planning a quantum leap.

It's no use saying "Tomorrow I will go for a new growth plateau." It doesn't work that way. You need at least three to six weeks of preparation. During this time, coax your muscles along, but be aware of holding back for the final push. Have no doubts that you will attain a new size and strength record. Be confident! Be absolutely certain that you will forge ahead, and you cannot possibly fail.

Eddie Moyzan, Gold's Gym, Venice

15

THE MUSCLE SLEEP

SNOOZING FOR SIZE

During World War II, the strategy and oratory of England's Winston Churchill united the British people against the tyranny of Hitler to a degree that no nation has been united since. Throughout the war, Churchill would stay awake and alert until three or four o'clock in the morning. His physical and mental endurance became a source of inspiration and wonderment to the Allies, and even Adolf Hitler began to wonder whether the British prime minister was mortal. Eventually, it was revealed that Churchill managed to stay alert into the wee hours of the morning because he took a quick nap in the afternoon.

There is little doubt that sleep is the most efficient way of maximizing recuperation. Experiments in Russia are said to have shown that several short naps during a 24-hour period can more than substitute for the uninterrupted seven or eight hours of sleep that most of us are accustomed to. Unfortunately, our jobs and our overall social system are not geared to this way of living. Few of us can work in three or four cat naps each day instead of the customary overnight slumber. Although people in Latin countries often enjoy an afternoon siesta—which recharges their batteries during the heat of the day, enabling them to be active in the relative cool of the evening—this is not the custom in most of Europe and North America.

The first person to use the term "muscle sleep" was British photographer and bodybuilding writer Chris Lund. As a professional photographer, Chris was frequently on call at all hours of the day and night, and the irregular sleep that this caused him was interfering with his training. Often he was not able to get to sleep until two or three in the morning, which left him pretty wiped out when it came time for his heavy-duty workout. So, he would try to make up for his lost sleep by taking a quick nap after eating his midday meal. Amazingly, Lund found that even though he had missed several hours of sleep the previous night, he could make up for this loss, or so it seemed, by grabbing a nap that lasted no more than 20 minutes. The muscle sleep had been discovered! It is the pause that refreshes.

Today, many of the pro bodybuilders have a light meal followed by a nap after their workouts. A few train twice a day and take two naps.

Some people, though not all, seem to have the ability to fall asleep at the drop of a hat. I will never

(Right) Serge Nubret
(Left) Alq Gurley

Andreas Munzer

forget the star high-jumper I saw some time back at an athletic event. He was an African American, 6 feet 8 inches tall, and his leaps seemed to be charged with atomic energy, as he gradually and systematically destroyed the opposition. After each jump, he would get back into his sweat suit, and then nonchalantly roll over and fall asleep! At the time, I thought this was highly amusing. I was amazed at how a man, with tens of thousands of people watching him, could fall asleep like that, then, upon awakening, seem to need little more than a couple of stretches to ready himself for a further attempt at the high-jump bar. Needless to say, he won the competition outright.

The next time I saw an athlete sleeping in between physical exertions was when I watched Serge Nubret train the last few weeks before he entered a NABBA Mr. Universe contest (which he won). After performing a dozen sets of high-repetition bench presses, Serge maintained a supine position, closed his eyes, and dozed off for a couple of minutes. Later on, after resuming his workout, Nubret again fell asleep after completing a series of leg presses. I asked him to explain his constant naps, and he said that during his last few weeks of preparation, he was on a very low carbohydrate diet that gave him limited energy. He felt that the naps enabled him to let go totally and allowed his body to recharge itself for continuing the workout.

In all honesty, the muscle sleep is neither practical for everyone nor necessarily recommended. I mention it for no other reason than that it is an option you might consider, used sometimes by pro bodybuilders during their last month before an important contest. If your lifestyle allows you the luxury of muscle sleep during a particularly grueling series of workouts, then some experimentation along these lines may help your progress.

At first, you may feel groggy after a 40-minute muscle sleep, and even confused with regard to your surroundings or the time of day. But when you get used to it, the muscle sleep will become a natural part of your day, and you will wake from it totally refreshed and ready to go. If you have had a full night's sleep, don't nap for more than 40 minutes; that would actually sap your energy and make you feel lazy by taking you too far into the sleep cycle. Needless to say, the most effective muscle sleep will be one that you take in a darkened room with as little disturbing noise as possible.

Milos Sarcev, Mike Matarazzo, Jeff Poulin, and Bruce Patterson

16
SHOULDERS

BUILDING BARN-DOOR WIDTH

Mother Nature seldom dishes out everything. In the case of shoulders, she either gives us narrow clavicles (the bone width) with plenty of cellular tissue to build big delts (shoulder muscles) or wide clavicles with a correspondingly poor allocation of muscle cells. Needless to say, we all want wide shoulders as well as trillions of muscle cells, so that we can really max out our shoulder impressiveness.

The shoulders are a complex area. Unlike the knee joint, which is one-directional (it only goes up and down), the shoulder has a ball-and-socket type of joint, which allows you to move your arm around in a circle with a very wide range of motion. In order to cope with this range, the shoulder muscles are divided into three separate "heads": the anterior (front), medial (side), and posterior (rear) deltoid.

No exercise really works all three sections at once, although there is some assistance from other parts of the shoulder, but it is practical to work one head at a time. The press-behind-neck works the side deltoid mainly, with some help from the rear head. The bench press, on the other hand, works the front deltoid, with a little involvement from the side deltoid—and, of course, the triceps and pecs.

Deltoid-isolation exercises are used frequently. They include the alternate front dumbbell raise for the anterior head, the lateral raise for the medial head, and the bent-over-flying movement for the posterior head.

Good as these isolation exercises are for hitting precise sections of the shoulder muscles, some kind of combination movement such as pressing is necessary for full deltoid development. The standard barbell press and the press-behind-neck are regulars among top bodybuilders. Other variations include dumbbell pressing, either together or in an alternating fashion. You may do these either standing or sitting. Sitting enforces greater exercise strictness, so this will probably benefit your deltoid development more.

Upright rowing is another recommended shoulder movement that can really balloon out your delts. Of course, strands and pulleys, though limited in overall body application, also help shoulder development. You can perform lateral raises from a variety of angles with both types of apparatus, and there is a very good shoulder movement exclusive to the strand-puller: the back press.

(Left) Garard Dente. (Right) Kevin Levrone performs a three-quarter back pose.

The competitive bodybuilder should realize the importance of large, fully developed shoulders, because this area cannot be hidden. Deltoids are seen from all angles—front, back, and sides. They are evident especially when you perform the double-biceps pose from the rear. To produce championship-quality shoulders, you need a routine that works all three deltoid heads. But should your time constraints limit you to performing just one exercise for the shoulder region, I would suggest either the press-behind-neck or the alternate dumbbell press. Alternating with a seesaw action is usually preferable, as the mechanics of the movement dictate that you don't lean back as you would tend to if you were pressing two dumbbells simultaneously. (By leaning back, the stress of the movement is thrown from the side deltoid head to the frontal area, so that you will get thicker delts but not wider ones.)

Wide-grip upright row (start)

press or the press-behind-neck) plus the bent-over-flying movement and the lateral raise. And advanced bodybuilders, who will be splitting their routine invariably into two or more sections, probably will find these most advantageous: two combination delt movements (such as press-behind-neck and upright rowing) followed by alternate forward dumbbell raise, lateral raise, and bent-over flying.

What follow are descriptions of some of the best shoulder exercises I have ever found. There is more than one way to skin the proverbial cat, but I believe my way is the best—although, in some cases, only by a small margin. The press-behind-neck exercise, for example, can be performed very effectively standing up, especially if the weight is taken from squat stands. But I have specified that it be performed sitting down, because I think it is somewhat more beneficial this way. This is not to say that you shouldn't experiment. Bear in mind, too, that an inferior exercise may prove an effective substitute for an excellent movement merely because your muscles (and mind) crave a change. As mentioned previously, a change can work wonders.

The tradition of mammoth shoulders has been with us for centuries. The ancient Greeks, for instance, revered their wide-shouldered, athletic Adonis, and popular fiction rarely introduces us to a hero who is not broad-shouldered. For the aspiring bodybuilder, the shoulder muscles are all-important. He needs them in virtually every pose, and when he is compared in the relaxed position, they are everything.

Most comprehensive workouts include two or three deltoid exercises. However, you should be aware that bench and incline presses work the frontal deltoids very strongly, whereas bent-over-rowing exercises work the rear deltoids strongly. Therefore, if you include some form of bench-press and rowing exercise, then you don't necessarily have to include isolation exercises for the front and rear deltoids. Certainly, there is no need to do so if you are a beginner or on a tight schedule.

Beginners should perform only one shoulder exercise: the press-behind-neck or the alternate dumbbell press. Intermediates will find it best to perform one combination shoulder exercise (such as the seated

Wide-grip upright row (finish)

Press-Behind-Neck
Side deltoid (6–12 reps each set)

Sitting down on a special upright bench with supports, take a loaded barbell, with your hand spacing sufficiently wide that when your upper arms are parallel to the floor your forearms are in a vertical position. Lower the weight as far as possible behind the neck, and immediately raise it when it touches your trapezius. Do not bounce the bar from your shoulders. Keep your elbows as far back as possible throughout the movement. Lock out the elbows as the arms extend overhead, but do not hold the position. Continue pressing up and down rhythmically without any pause.

Alternate Dumbbell Press
Side deltoid (8–12 reps each set)

Work in sitting position, your back flat. Start the movement with a pair of dumbbells held at the shoulders, palms of your hands facing inward or forward. Hold your elbows back to maintain stress on the side deltoids. Start with your weakest hand, and alternately press first one dumbbell, then the other, in a seesaw fashion. Lock out the arm each time you press the weight, but do not maintain the straight-arm position; as soon as your arm straightens, lower it, and continue the exercise with no pauses whatsoever.

Upright Rowing
Front and side deltoids (8–15 reps each set)

Use a fairly wide grip. The wider the grip, the more stress is put on the side deltoid. (A narrow grip will put more stress on the front deltoid and the trapezius.) Always straighten your arms at the bottom of the exercise, and start your pull slowly, gathering momentum as the weight rises to your chin. Keep the up-down movement rhythmic. Maintain an upright stance with your feet comfortably apart (12 to 15 inches). As the bar rises, try to keep your elbows as high as possible.

Lateral Raise

Side deltoids (8–12 reps each set)

There are a thousand and one ways to perform this exercise, all with one aim: to throw the stress onto the side deltoids. (Everyone wants more shoulder width!) Perform this exercise standing or seated on the end of a bench. Your feet should be together and flat on the floor, ankles touching. The arms must be bent to almost right angles to throw stress on the all-important lateral deltoids. Raise the weights, from the arms-straight-at-sides position to level with your head, and immediately lower them. Keep your palms facing downward throughout. At the point of extreme effort, try to lean forward into the exercise, rather than backward (which will put stress on the powerful front deltoids).

The Arnold Press

Side and front deltoids (10–15 reps each set)

This exercise is a pure bodybuilding movement. The mechanics involved are not at all in line with the accepted or "natural" way to lift weight overhead, but the movement does build muscle. Reportedly, the exercise first was used during the "Golden Age" of bodybuilding, the 1960s at Muscle Beach. It was a favorite with Larry Scott, who is said to have explained it to the great Arnold Schwarzenegger. Arnold then used it so successfully that it became known as the "Arnold Press."

Begin the movement by holding a pair of dumbbells as you would at the top of a dumbbell curl. From this somewhat irregular position, press the dumbbells upward, while rotating your thumbs inward. Do not lock out the arms at the conclusion of the part-press, part-lateral-raise exercise, but continue the up-down movement without pause to exhaust the deltoids fully.

Single-Arm Lateral Raise with Dumbbell
Side deltoids (8–15 reps each set)

A great favorite with old-timers Boyer Coe and Larry Scott. The beauty of this exercise is that there is little stress on the lower back (which can cause discomfort) and you can lock yourself into a specific position, thereby avoiding all possibility of cheating.

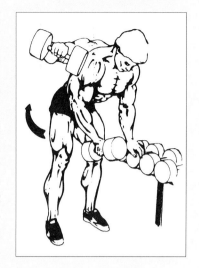

Place your left arm on a suitable support (about 30 inches high). Most trainers use a dumbbell rack or low table. Standing with your legs apart, hold a dumbbell in your right hand. Adopt a comfortable position with your torso bent forward at an angle of 70 to 80 degrees. Raise the dumbbell out to the side, keeping the palm of your hand facing downward. Concentrate on making the shoulder muscle lift the weight (the arm should be unlocked but not excessively bent). Don't start this exercise with maximum thrust, otherwise you just lift the bell with momentum. You don't see this exercise variation used much these days, but it is very effective.

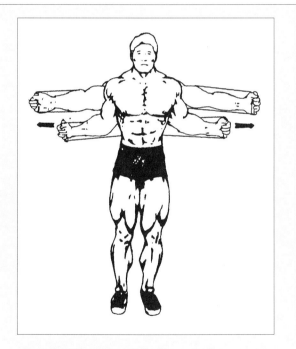

Back Press with Strands
Side deltoids (15–25 reps each set)

Start by holding a set of strands behind your back, palms facing forward, elbows tight in at your sides. Press out the arms to a straight crucifix position. As soon as the arms straighten, return them to the original position and repeat the process. Rubber strands work better than steel springs.

Seated Dumbbell Press
Side deltoids (6–12 reps each set)

While seated, hold two dumbbells at your shoulders. Keep your back straight and your head up. Press both dumbbells simultaneously to the overhead position. Do not lean back during the exercise. Lower and repeat with a steady rhythm.

Bent-over Flying
Rear deltoids (10–15 reps each set)

This important exercise is one of the few that work the rear shoulder region almost exclusively. Full development in the rear delts is a must in today's competitive arena, because anything less than maximum posterior delt size displayed in a back pose will be cause for criticism. The back deltoids also contribute to side-view thickness and counteract any appearance of round shoulders you may have.

Sit on the end of an exercise bench with your knees and feet together. Lean forward until your chest touches your thighs. Raise your heels off the floor, so that your thighs and chest are locked together, supported by your toes on the ground.

Holding a pair of light dumbbells, raise your arms out sideways, palms facing each other, until the weights are as high as you can get them. The arms should be un-

locked (not straight) to alleviate pressure on the elbows. Always start this movement slowly, forcing your rear deltoids to work with strength to build up momentum during the last 12 inches before the conclusion of the raising motion. (If you start off quickly from the hang position, you minimize the rear-deltoid involvement and kill the main usefulness of the exercise.)

Alternate Dumbbell Front Raise
Front deltoids (10–15 reps each set)

Hold a pair of dumbbells while in the standing position. Raise one hand directly forward until it is well above your eyes, and then, as you lower it, raise the other hand. Continue this seesaw motion without pausing in any position. (Do not lift both dumbbells out in front at the same time, because this would cause the torso to adjust the balance by leaning backward, thus negating some of the stress on the front deltoids.)

Lateral Raise with Strands or Pulleys
Side deltoids (15–25 reps each set)

A good finishing or pumping exercise to conclude your deltoid workout. You may use two pairs of strands, one hooked under each foot (or use low pulleys if you train in a well-equipped gym). Raise and lower your arms at equal speed without any pause. Hold your torso bent forward slightly in order to put stress exclusively on the lateral head. Make this a continuous tension movement by keeping continuous pressure on the deltoids

(Above) Mike Matarazzo
and Porter Cottrell
compare triceps.
(Right) Paul Dillett

17
CHEST
SCULPTING THE PECS

M en like Nasser El Sonbaty get an enormous amount of pectoral development from the bench press. Although this exercise is no longer the all-time favorite of the competitive bodybuilder that it used to be, it works for most people. The regular bench press is 80 percent front deltoid, but to my mind it is still an excellent chest exercise.

Perhaps it gives the most gain to those who have a fairly shallow rib cage. With them, the bar has a great distance to travel and thus gives a good stretch to the pectorals. On the other hand, a barrel-chested person, because of the size of his chest, may not be able to lower the bar very far, although that is not always the case. Former Mr. Universe Reg Park had a fairly thick rib cage and still got a great deal from the bench press. Even if you do have a thick or deep chest, you can always increase the stretch by lowering the bar to your neck. (No bouncing, though!) Man being ever the inventor, there is also a cambered bar on the market now that allows even the most deep-chested of men to get a full and complete stretch of the pectoral muscles.

The beauty of the bench press is that it is performed from a very comfortable and stable position—lying on your back, faceup. After you get used to it, you do not have to concern yourself with balance or performance difficulty. As a result, and because the belly of the pectorals and triceps is involved, your strength and development grow when you practice the bench press regularly.

Many people think I am against the bench press, that I have an ongoing vendetta against this movement. What I am against is the abuse of the exercise. Fortunately, it is not so widespread today, but in the sixties and seventies some bodybuilders were spending up to two-thirds of their workouts bench pressing! As a result, we saw bodybuilders with balloons instead of pectorals. What made matters worse was that these men were not giving enough time to their shoulder training, so the rest of them didn't have the proportions to complement their pectorals. They had no delts, no width. In fact, they looked like a deformed species, even though they undoubtedly felt like heroes. At the very least, big pecs should be accompanied by big deltoids.

(Above right) Here's an early pic of Reg Park soon after he won the Mr. Britain title at the age of 21. (Left) Paul-Jean Guillaume

John Simmons, Detroit

ceps are also activated strongly by narrow-grip bench presses.)

As you lower the weight, you activate whatever area is in line with the bar. If you lower the bar to your lower chest, you will work the lower chest. Bring the weight to the middle of your pecs, and that is where you will stimulate most growth. Lower the bar to the upper chest for . . . right! . . . upper-pec development. Naturally, there is some spillover effect. Even though you are working for development in one area, bear in mind that other parts of the pectorals will also be stimulated.

When sculpting your pectorals, you should be aware that incline presses with a barbell or dumbbells will work the upper chest. Flyes work the outer pecs (some inner development is achieved when the dumbbells are brought together, but at this point the resistance is minimal). Regular dips involve the lower pectorals, but if the dipping bars are moved out to 28–34 inches (taller people need wider bars), then you will work the upper and outer area of the pectorals.

Pullovers help the rib cage, but do not expect any dramatic rib cage expansion. Expansion will eventually take place, but only within the framework of your skeletal genetics.

Watch It!

As with other exercises, forget about hoisting up the weight. Bouncing, twisting, lifting the hips from the bench, in order to raise the weight, are not the best ways to build great pecs. On the contrary, you should use the weight correctly as a tool to achieve your goal. As Arnold Schwarzenegger says, "The trick to bodybuilding is to put an overload on your muscles. The secret is not so much to get the weight up as it is to push up a heavy weight with the isolated strength of the muscles you are trying to train."

One of the greatest errors in training the chest, according to expert Johnny Fitness, is lack of concentration. It's essential to flex the pectoral muscles throughout the movements. Another mistake is to follow someone else's routine, set for set, without concern for the particular needs of your own body. Finally, always remember to stretch the pectorals fully. After your first warm-up sets, you can really bring out the arms and fully extend the motion affecting the pectorals. In most cases, the use of dumbbells allows for more of a stretch than the use of barbells.

These are the best chest exercises.

Although the bench press can build a fairly good all-around chest, you should be aware of the other chest exercises that can help to balance your chest development. The bench press may not do it all for you. In fact, it probably won't. You will be rewarded, however, if you vary the width of your hand spacing when you bench-press. A wide hand spacing puts the exercise stress on the outer part of the pectoral, a medium grip will hit the middle part of the chest, and a narrow grip will work the inner pectorals. (Of course, the tri-

Bench Press
Overall pectoral area (5–15 reps each set)

The bench press can be tailored to any part of the pectoral muscles. If the exercise is to benefit the upper chest, then lower the bar to the neck. If you lower the bar to the middle of the pectorals, that is where the effect will be felt. Narrow-grip bench presses will activate the inner pecs. The outer chest is worked with a wide hand spacing. You can tailor the bench press to pinpoint any area for muscle development by choosing the right grip width and bar groove.

The standard way of performing the bench press is to take a grip with your thumbs about 3 feet (90 cm) apart, which allows the forearms to be vertical when the upper arms are parallel to the floor.

Start while lying faceup on a bench. Take a balanced hand placing, using the thumbs-under-the-bar grip (not essential). Lower the weight from the arms-straight position to the pectorals. Touch the bar lightly to the chest (no bouncing) and press upward. Keep your elbows under the bar, and don't allow them to come close to your body.

Beginners may find that the bar starts to fall either forward or backward, or that the weight is rising unevenly because one arm is stronger than the other. After a few weeks, however, you won't even have to think about balancing the weight or lowering it without wobbling, because you will have then developed a perfect groove.

When you lower the bar to the chest, don't allow it to drop! Always control its descent deliberately, especially if it is a heavy weight. Control its downward path, and you are assured of a positive upward movement.

Parallel-Bar Dips
(8–20 reps each set)

A wonderful chest movement, especially if the bars are set fairly wide apart (28–34 inches, or 70–85 cm). Narrowly set parallel bars will promote more triceps activity, but will still work the lower and outer pectorals. Wider-set parallel bars will benefit the upper-outer part of the chest. This development will make you look wide in the upper torso and shoulders.

When working the chest on the dip bars, place your legs in front of your body, keeping your head down (chin on chest) and elbows well out to the sides. Lower yourself as far down as possible, and lock out the elbows as you straighten up.

97

Chris Cormier performs the dip exercise at the open-air pen in Venice, California.

Incline Dumbbell Bench Press
Upper chest (8–12 reps each set)

Start by lying on an incline bench set at a 35 to 40 degree angle. (More than 40 degrees will put too much emphasis on the front deltoids.) Press the dumbbells simultaneously straight upward, lock out the elbows, and immediately lower the weights to the starting position. Keep the up-down movement going without pause. Palms should be facing forward throughout the exercise.

Incline Flyes
Upper and outer chest (8–12 reps)

Adopt a secure position on an inclined bench (a 30 to 40 degree angle is best). Hold up a pair of light dumbbells, then allow your arms to lower slowly out to the sides. Keep elbows bent slightly throughout the exercise. Raise and lower slowly, keeping the weights under control as each repetition stretches the chest.

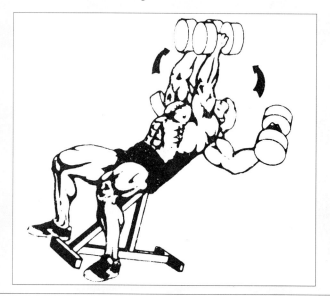

Flat Bench Press with Dumbbells
Mid-chest area (6–12 reps)

Lie on your back, and hold two dumbbells with your arms fully extended at right angles to the floor. Lower both dumbbells to your chest, and immediately press them up again to their original position. Do not bounce the weights from the chest. Keep your elbows out from the body during the movement.

Supine Flying

Outer pectorals (10–12 reps each set)

Years ago this exercise was performed very rigidly with light weights on the floor. Very light dumbbells were used because the experts at that time insisted that the arms be fixed in an elbows-locked, straight position. Today, we still insist on a fixed position, but one in which the arms are bent as though they were in a plastic cast. This takes the strain off the elbow joint, allows more weight to be used, gives you greater dumbbell control, and . . . yes, bigger chest muscles!

While lying faceup on a bench, with your feet planted firmly on the ground, lower and raise the dumbbells out to the side. Really go for the stretch once your muscles are warmed up. Keep the arms "locked" in the unlocked position.

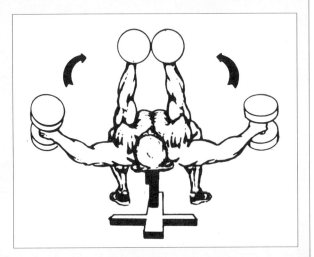

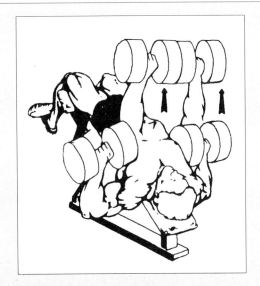

Decline Bench Press with Dumbbells

Lower chest (8–12 reps)

Lie on a declined bench, as shown. Press up and then lower both dumbbells simultaneously, as you would in the dumbbell bench press on a regular flat bench. Keep your elbows as far out (sideways) from the body as possible.

Incline Barbell Bench Press

Upper chest (8–12 reps)

Take a loaded barbell while in an incline position on a secure bench. Lower it slowly to the upper chest, your elbows out to the side, and then push it to the arms-straight position. Lower and repeat.

100

Cable Crossovers
Overall chest (10–15 reps)

This is a specialized exercise. Cable-crossover machines are expensive, but most modern gyms do have them. Holding the cable handles, bend your arm slightly at the elbow and bring your hands together in front of your chest or hips. Try to concentrate the action on the chest area. Keep the arms "locked" in a slightly bent position throughout the exercise.

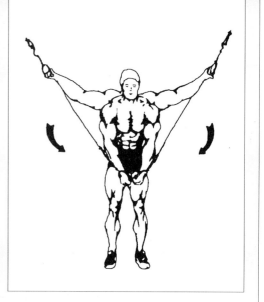

Incline bench press (start)

Incline bench press (finish)

18

ABDOMINAL TRAINING

WASTING AWAY FOR MIDSECTION IMPRESSIVENESS

You should do your best to never allow enough fat to accumulate so that the abs disappear. But should fat start to settle around your midsection, it will have to be removed at some point, which usually means that you will have to put yourself on a reduced-calorie diet for a considerable time. This will, of course, tend to hold back or diminish overall muscle gains. It's far better to keep your abs throughout your training career, so that when contest time approaches you will require only minimal dieting.

Many beginners are puzzled as to whether or not they can change the shape or evenness of the ab muscles. Of course, a set of "line-up" abs running across the tummy in even rows is most impressive, but few people actually possess this, not even the great Steve Reeves. Unfortunately, no amount of exercise will change the placement of your ab muscles. If they are uneven now, then that is how they will stay for life.

Famous bodybuilders who do have even rows of abdominals include old-timers Mohamed Makkawy, Leo Robert, Dennis Tinerino, and Samir Bannout. Among today's champs, we can include Milos Sarcev, Shawn Ray, Paul Dillett, Francis Benfatto, and Eddie Robinson. However, it is no great sin to have uneven layers of waist musculature. Mr. Abs himself, Irvin "Zabo" Koszewski, had uneven abdominals, yet he won just about every "best abs" contest he entered until he was well past 40 years of age. Everybody, but everybody wanted abs like Irvin's.

Among the uninitiated, it is commonly believed that one can get the abdominal muscles to show by performing a few sets of sit-ups or leg raises every day. This is not true. If there is a covering of fat around the waistline, then the direct abdominal exercise only will build the muscle density under the fat. Invariably, this is not enough to get the abdominal muscles to show up as impressively as one would like. In fact, it is quite likely they won't show up at all.

The answer, then, is diet. You simply must revise your nutritional intake so that you ingest fewer calories. Remember, if it took a year or two to put the fat on your waist, you are not likely to remove it with a couple of weeks of dieting. A very determined and sometimes lengthy period of dieting often is required.

(Right) Mohamed Makkawy
(Left) Abs to die for—Paul Dillett

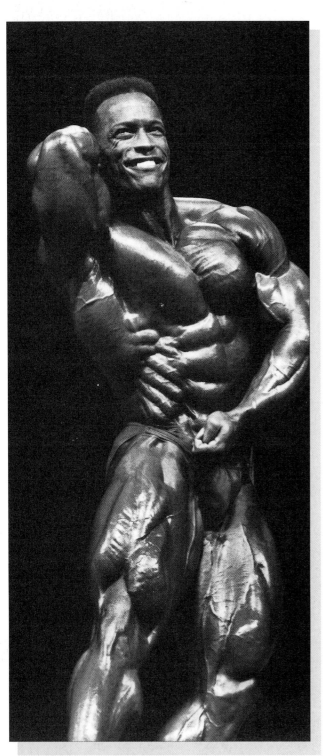

Look at the incredible "intercostals" of IFBB pro Shawn Ray.

It is a biological fact that 3,500 calories equal one pound of fat. Accordingly, if you wish to lose one pound of fat every week, you would have to cut 500 calories per day from your normal diet. However, it is easy to lose one pound a week by making small food sacrifices here and there.

For many years, bodybuilders trained their abs with very high repetitions, although this practice tends to be frowned upon by modern bodybuilders. Veteran Bill Pearl, when training for a contest or exhibition, has been known to perform hundreds of reps for the waist. Frank Zane, another old-timer, performed sets of 200 reps. Irvin Koszewski trains his waist by the clock: 30 minutes per set at minimum. In his heyday, he never did less than 1,000 repetitions.

MuscleMag International was one of the first bodybuilding magazines to call for rep moderation in working the abs. Trainer Mike Mentzer followed up by advocating a system of no more than 12 to 15 reps for abdominal movements.

Both methods work, high reps and moderate, but today saving time is often an important factor. Why do more when you can get the same results by doing fewer?

Actually, the abdominal area is very sensitive to heavy exercise, and high-intensity effort is suitable only for the more rugged types. If the average man were to go all-out with heavy reps with his abdominal exercises, this could stop his overall gains. The midsection fascia is the center of the body's nervous pathways. By overworking it, you could shock your entire system, which could cause a temporary shutdown of the growth process.

My own thoughts run to performing moderately high repetitions for the abs: around 15 to 30 per set. You will find out by trial and error what works best for you. In any case, it is not a good idea to do a great deal of abdominal exercising late at night. Such overstimulation could make it difficult for you to fall asleep.

Have you ever noticed, when looking at some of the muscle magazines, that many bodybuilders have oversized, chunky, and noticeably bloated stomach muscles? This usually is caused by taking anabolic steroids, often in large, uncontrolled quantities. Growth hormone, a drug taken by numerous Olympic athletes and bodybuilders, also causes excessive stomach bloat.

Having an unaesthetic midsection bloat will not win over a contest audience or the judges. If your abdominals look a little bloated as a contest date is

Even in this "relaxed" pose, Milos Sarcev looks unbelievable.

zooming up, it may be a good idea to cease all direct abdominal work during the last 10 days. You can keep them toned by your posing practice, which you will undoubtedly pursue with vigor as the show date draws near. Without heavy exercise, the edema (bloat) should vanish and make your waistline look much trimmer. Please note that steroid-induced bloat or waist thickness caused by the use of growth hormone cannot be reversed easily.

I have always admired those bodybuilders who combine exercise and diet to bring out their abdominals, especially when they manage to bring them out from nowhere (I should say from behind a wall of flab). In some cases, this happens over and over again. Yet, you should not make a habit of losing your abs at the end of each summer only to blitz off the flab the following spring.

Special care should be taken not to stretch the abdominal wall. This can occur through overindulging in beer (especially at one sitting) or through performing heavy squats without wearing a strong leather belt. Look at the abdominal wall as a coil spring. Pull it out a certain distance and let go. It zips back into the coiled position. But pull it out a greater distance, and the strain is too much for the metal. It doesn't spring back when you let it go. Nothing you can do will make it zip back.

Today, many bodybuilders have superb abdominal muscles. This is in contrast to the physique stars of the past. Only a handful of them had outstanding midsections.

There are scores of different midsection exercises, and they all work to a degree. Here are some of the best I have found.

Incline Twisting Sit-ups

Lie back on an incline board set at any angle you choose (the steeper the angle, the lower the part of the waist is worked). Your feet should be held to the board with a strap (or bar under which the feet fit). Place your hands behind your head, and curl upward. Keep the knees bent slightly throughout the movement.

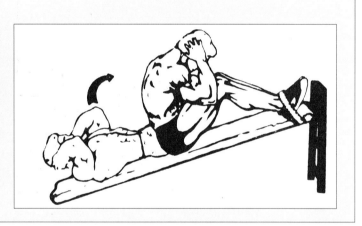

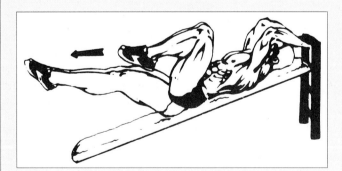

Incline Knee Raise

Lie back on an incline bench (the angle can be varied), and secure your position by holding on at a suitable place. Raise your legs, bending the knees as they rise. Lower to the straight-leg position slowly, and repeat.

Hanging Leg Raise

Hang from an overhead horizontal bar, with your arms about 30 inches (75 cm) apart. Keeping your legs straight, raise them until they are just past the parallel-to-floor position; then lower and repeat. Try not to let the body build up a swinging motion. This exercise works the very important lower-abdominal region, right down to the groin.

For those who are unable to perform this exercise with straight legs, start off with the knees bent. Tuck your knees into the waist at each repetition, and point your toes downward. Start the raise slowly, with positively no swinging. After a few weeks, you will be able to graduate to the straight-leg style.

Inversion Boot Sit-ups

An excellent movement for the lower tummy area. Using a pair of inversion (or gravity) boots, hang from a strongly supported horizontal bar. Curl upward in an upside-down sit-up, lower slowly, and repeat. Gravity boots are available from specialized exercise stores.

Roman Chair Sit-ups

You need a Roman chair to anchor your legs in position and allow the trunk to sink below parallel, thus working the abdominal region to a greater degree. Perform this with a steady rhythm and no bouncing. This is the favorite exercise of Irvin "Zabo" Koszewski, who has won more "best abdominals" awards than any other bodybuilder. Arnold Schwarzenegger also likes this movement

19
PUTTING YOUR BACK INTO IT

WORKING THE LATS AND TRAPS

M ale or female, one of the most impressive areas of the body is the back. The human body is arranged in such a way that the two biggest muscles of the back— the lats (latissimus dorsi) and the traps (trapezius)—actually can be seen not only from the back, which is expected, but also from the front. This is true especially when these two major back muscles are developed fully. The other area that concerns the bodybuilder is the lower back—the two columns of the lumbar region, the erector spinae. These are very powerful muscles that, when fully developed, create an impression of strength and virility.

Beautifully sculpted backs, as we are used to seeing them today, just were not around 50 years ago. True, the muscle men back then did have thickness and mass, but the enormous, excessively tapered backs of our current era are a relatively new phenomenon.

The supreme back of the forties belonged to John Carl Grimek, followed by the perfectly proportioned back of the Hercules of the screen, Steve Reeves. This man had an absolutely amazing back, developed in every facet. His traps reached up into his columnlike neck. He had wide, flaring lats, and his lumbar region also was superb. It's little wonder that Reeves won every top title of his era. Reg Park took back development a step further. A new massiveness came about in the fifties, and for more than a decade Reg was ahead of the pack. Like Reeves, he had width and taper, but with an added dimension: Herculean thickness! In the late sixties and the seventies, two bodybuilders vied for the top titles: Sergio Oliva and Arnold Schwarzenegger. Both had amazing backs. Sergio had the edge over Arnold when it came to V-shape (will his phenomenal taper ever be equaled?), but Arnold had it over Sergio with regard to thickness and etched-in muscularity.

Today, we have men with even more incredible backs. We have the rock-hard detail of Shawn Ray. His back, when flexed, more approximates a den of serpents than anything anthropoidal or from the family of Homo sapiens. Mr. Olympia Dorian Yates has a back so wide that it defies description. Prior to this era, the widest back belonged to IFBB pro Tony Pearson. But today Yates also has the added ingredient of Godzilla-like thickness. In addition, we have men such as Ronnie

(Right) Super-cut Hamdullah Aykutlu
(Left) What a back! Flex Wheeler

Top IFBB pro Ronnie Coleman

Coleman and Aaron Baker. Both have amazingly muscled backs, especially when viewed in the rear double-biceps pose.

Some men worry that the development of the trapezius will detract from the visual width of the physique. This is not so. What detracts from the visual width of the body is underdeveloped shoulders and wide waist and hips. The only time you should refrain from performing specific trapezius exercises is if you have inherited a short neck. Heavy trap development will give the short-necked bodybuilder an unattractive hunchiness.

One of the most spectacular traps that I have ever seen belongs to veteran bodybuilder Serge Nubret, who incidentally seldom works them directly. Because of his superbly small waistline and hips, and even though he has huge traps, Nubret looks extremely wide in the shoulders.

The lats are the biggest muscles of the back. They are "wings" that can be seen from the front under the arms. There are two ways to approach lat building. First, the scapulas (shoulder blades) must be stretched out. This is accomplished with wide-grip chins or wide-grip machine pulldowns, either in front or behind the neck, or by doing the lat-spread pose. Then thickness must be built in the area, usually by performing one or more of the various bent-over or seated rowing movements.

As a result of inherited traits, some men possess high lats and others low lats. The vast majority of us are somewhere in between. It is generally accepted that a man with high-lat development should perform plenty of rowing movements, pulling the bar into the waist to work the belly of the muscle. A low-lat individual, on the other hand, requires no further development in the bottom regions and could concentrate on stretching out the upper areas with wide-grip chins. Natural shape or type cannot be turned around completely, of course, but some changes can be accomplished.

It is important for most bodybuilders to train the lats for width and thickness. This means that they should make a point of performing at least two lat exercises—one stretching movement and one rowing movement—during each back workout.

The question has been raised as to which is the superior exercise for pulling out the lats: the wide-grip chin or the wide-grip lat-machine pulldown.

A lat spread from Britain's Dorian Yates

Theoretically, the lat machine wins, because the pulldown motion can be controlled extensively, in that you can bring the bar way down below the shoulder level if you wish (by adding to the range of resistance). Also, the lat machine permits a greater variety of reps with very little inconvenience. If you wanted to perform numerous sets of 30 reps, for example, you would have to use a lat machine—unless you are a superman.

However, it must be said that the chin, in which the body is pulled up to the bar, does confer some benefits that the pulldown motion does not. This is my feeling, and apparently that of most top bodybuilders. In comparison to the lat-machine pulldown, the chin pulls out the lats more dramatically and adds more to the thickness of the upper back.

Here are my recommendations.

Bent-over Rowing

This is one of the most popular exercises for putting some meat on your lats. Grab a barbell with your hands about 24 inches (60 cm) apart. Bend your knees slightly, and then keep your head as high as possible, while bending your torso parallel to the floor. Keep your lower back flat, your seat stuck outward, and pull up vigorously on the bar. Pull it into the tummy, not the chest. Lower it until your arms are completely stretched, and more. Do not rest the weight on the floor until the set is completed. Pull up, and repeat.

Low Pulley Rowing

Perform this movement with a long cable machine. Secure your feet against the apparatus, and pull the cable handles horizontally into your midsection. Hold for a second, and slowly allow your arms to straighten and ultimately stretch your lats. Pull in again, and repeat. Aim to maximize that stretch as the arms straighten.

T-Bar Rowing

This movement, primarily for the belly of the latissimus, is performed on a special apparatus. The movement is almost identical to bent-over barbell rowing, except that one end of your lever bar is anchored to the floor or the machine. As a result of this, there may be less strain on the lower back.

Wide-Grip Chin

Grasp an overhead bar using an overgrip (palms down) at least a foot wider than your shoulders on either side. If your shoulders are 2 feet (60 cm) across, take a grip about 4 feet (120 cm) wide. Pull upward, keeping your elbows back throughout the movement. You may pull up so that the bar is either in front or behind your neck. That choice is entirely up to you. Some bodybuilders like to change around for variety, but it would not be correct to say that one form is superior to the other. Lower until your arms are straight, and repeat.

Once you can perform 12 to 15 reps, it is a good idea to attach added weight with the aid of a weight belt. After that, you can build up your reps again.

Single-Arm Dumbbell Rowing

This is a "total" lat exercise, but one without lower-back strain, because your nonexercising arm is used to support the entire upper body. Pull the dumbbell up into the midsection, and lower it until the arm is extended all the way down. Then try a little harder to lower it even more. Maximize the stretch.

Lat-Machine Pulldowns

This exercise has to be performed on a lat machine. Take a wide overgrip on the bar, and pull down as far as you can. Though not as effective as the wide-grip chinning exercise, this one does have the advantage that you can use less resistance and therefore pull the bar lower and work your lats over a greater range of movement. You may pull to the front or the back of the neck.

Good Mornings

Stand with your legs set comfortably either together or apart, a loaded barbell across your shoulders. Keeping your back flat, bend forward at the waist and straighten up. Hold your head as high as you can throughout the movement. Do not use heavy weights until you have fully mastered this movement.

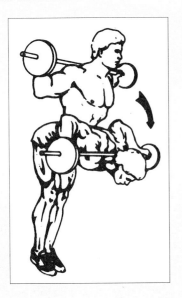

Shrugs

Hold a barbell with a shoulders-width hand spacing, while standing upright. The bar may be held in front of the body or behind it, as you prefer, and you may also substitute dumbbells for the barbell. Keeping the arms locked, raise the shoulders upward toward the ears, as high as possible. Then rotate them backward and down. Do not bend the knees. Concentrate on the up-down rotation of your shoulders. Some bodybuilders use a Universal bench-press machine instead of free weights for this exercise. The shrug is considered to be the best all-around trapezius-building movement.

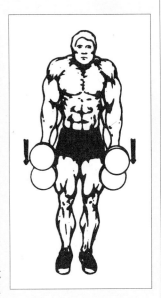

Prone Hyperextensions

This is performed on an exercise unit especially designed for the movement. Place the legs and hips front downward on the bench. The upper body should be free to rise up and down. Place your hands behind your head, and lower your trunk toward the floor. Rise until your body is in a straight line. Lower, and repeat.

In time, as your lumbar region strengthens, you may hold a barbell behind your head, as in good mornings. This movement is a great favorite with bodybuilders today because it helps guard against lower-back injury.

Jay Cutler performs the wide-grip chin (start).

Wide-grip chin (finish)

20

QUAD TRAINING

MAXING OUT THE THE UPPER LEGS

With the advent of new machines and techniques, the competitive bodybuilders of today exhibit entirely different quads from those we saw in the sixties or before. In those days, cross-striations were unheard of. Only the late Vince Gironda had the upper-leg separation and detail that is so important for winning today's contests. It would be true to say, however, that then, as now, the regular back squat was "king."

Real thigh impressiveness cannot be built without squats, except perhaps by the most genetically gifted of men. If you are looking for more thigh size, then you must base your workout around the regular back squat. The other upper-leg exercises, though important, are of a more supplemental nature.

Squatting is a very natural exercise, but it was not used by weight men extensively until the introduction of squat stands. Prior to that, lifters would shoulder the weight after first standing it on end, and then they shuffled it into position on their shoulders.

One of the first advocates of the barbell squat was the Milo course, published at the beginning of the twentieth century. At that time, heavy weights were not recommended. Alan Calvert, who wrote the course, recommended a double-progression system, which even today cannot be bettered for beginning bodybuilders. It recommended twice as many repetitions for the legs as for other body parts. (Tom Platz probably would agree with this!)

This is the way the Milo system worked: You took a set poundage (on the light side) for an exercise (say, the curl), and you proceeded to curl it 5 reps. You worked out every other day. On the second exercise day, you also did the movement 5 times. Then, on the third exercise day, you increased the reps to 6. After two more workout days, you went on to 7, and so on, up to 10 repetitions. When you had done 10 reps twice with the starting weight, you increased the weight of the bar by 5 pounds, and began all over with 5 reps. In doing leg exercises, you started with 10 reps and increased by 2 reps every third exercise day, until you reached 20. Then you increased the weight by 10 pounds and dropped back to 10 reps.

This system has several marked virtues. One is that the body gets used to small amounts of progression, which definitely helps in avoiding sticking points.

(Right) Tom Platz
(Left) Paul Dillett's amazing legs

In order to handle more and more weight, many of today's aspiring bodybuilders pile on the discs to maximize intensity and strength—only to find that they are driving themselves quickly into a muscle-building stalemate. Another virtue of the Milo system is that it is cyclic in nature. When you reach the maximum number of repetitions, it is a tremendous relief to increase the weight by only 5 pounds (or 10 in the case of squats) and then go back to half the number of repetitions. It provides a sort of rest period (cycle training) in the first part of each series of progression. Even today, this system is great for the first five or six months of any newcomer's weight training. After that, it simply doesn't work.

The greatest proponent of the regular back squat is Tom Platz, who had phenomenal thigh development. According to Tom, most bodybuilders do not squat correctly. "I watch guys squat, and 80 percent of the time they don't do the movement right," he says. "Their feet are too far apart, their squat depth is too shallow, or they lean too far forward as they squat." Needless to say, when you lean too far forward, you put more stress on your buttocks than on your thighs.

Let's drop in on the magnificent Platz as he prepares to squat. He initially does some stretching. The hurdler's stretch for the front and back of the legs is first. After that, he does a hamstring stretch from a flat

The incredible legs of Tom Platz

David Palumbo goes all-out on his squats.

bench, keeping his legs locked tight. What strikes you immediately when you observe Platz is that he is totally prepared. He is wearing tight sweat pants, a regular leather lifting belt, plus a tight-fitting garment around his torso "to enhance the secure feeling that you must have to squat properly and successfully," he explains. His shoes are weightlifting boots with a raised heel. Platz doesn't recommend flat-footed squats in bare feet or tennis shoes.

He approaches the bar with a zeroed-in attitude. He means business. The bar, taken from the racks, is on his shoulders. A couple of steps back, and his feet are set—never wider than shoulders' width, and usually considerably closer. His toes are pointed slightly outward.

As Platz lowers into the squat, you know he is in total control. He holds his head high, and places his hands on the bar about halfway between his shoulders and the plates. His torso is flat-backed and upright. He lowers to parallel and beyond, and returns to the upright position. Even other professional bodybuilders stop what they are doing to watch Platz squat.

As he sinks slowly all the way down again, his knees travel out directly over his toes. He stays tight, keeping his back and torso muscles tensed and . . . up again! The breathing is heavy. He breathes in fully before squatting again. Out jets the air as he straightens up. After the set, his jellyfish thighs make him yearn to sit down, but he resists the temptation. He walks around, keeping the blood circulating, which helps him to recuperate quickly for the next set. To sit down would be encouraging workout lethargy. He may allow himself that luxury after completing his entire leg program.

Incidentally, Tom Platz had 20-inch thighs when he took up weight training. Later, at less than 200 pounds of body weight, he squatted with 600 pounds. He has also done an incredible 28 reps with 405 pounds, and 52 reps with 350. According to writer Bill Reynolds, prior to 1977 Platz was doing 10 straight minutes of squatting with 225 pounds!

The squat will help along your overall gains more than any other movement. You should beware, however, of doing too much. An excessive number of sets with maximum weights can lead to overtraining. It is seldom advisable to squat more than twice a week, and even then, one session should be somewhat watered down. In other words, don't shoot for two weekly squat workouts using the training-to-failure technique.

After squatting, you should employ (unless you are a beginner) at least one other frontal thigh exercise. The hack-machine squat is a good supplementary movement for added size and shape. The 45-degree leg press is also useful as a supplementary quad-building movement.

The leg extension is useless, at least as a quad builder. It is a very popular movement, however, and is

Daren Charles performs leg extensions while Jim Mentis looks on.

used by bodybuilders at all stages. Its charm may be that it is a delight to behold! There you are, your thighs held in place by a bench, your lower legs pumping up and down. It looks like something is working. But where's the size? The truth of the matter is that the leg extension is a fantastic knee exercise, and that's it!

When training the quads, pay attention to developing proportion. For instance, if you have just performed heavy quarter squats for the thighs, you want to guard against developing "turnip thighs" (large upper thighs, small lower thighs). In order to balance development, you now should work the lower thigh with some specialized movements, such as the sissy squat or hack lift. Because of its fixed, vertical-slide position, the Smith machine enables you to place your feet farther forward than you could in a free-weight squat with a barbell. Therefore, more stress is thrown on the entire thigh area, rather than on the hips and butt. Trainers who do not wish to enlarge their buttocks should opt for the Smith-machine squat, keeping their feet forward of the bar to minimize stress on the buttock muscles.

It often is asked whether the leg press is as effective an exercise as the regular back squat. Heavy weights can be hoisted using the leg press, and the apparatus is useful in working the thigh relatively comfortably from a variety of angles (you can alter foot placement dramatically). But as an overall-thigh exercise, the leg press is definitely inferior to the squat, which is truly the most effective quad builder currently known to mankind.

Here are the best leg exercises.

Ronnie Coleman squats for leg power and size.

The Squat
Entire thigh area (6–20 reps per set)

Take a weight from a pair of squat racks, and hold it, hands on bar, at the back of your neck. If needed, place your heels on a two-by-four block of wood to improve balance. Some people just cannot squat flat-footed. It forces them to adopt a very wide foot stance, and even so, they are forced to lean too far forward when squatting down.

Breathe in deeply before squatting down. Keep your back flat and your head up throughout the movement. Breathe out forcefully as you raise up.

The Hack Squat
Mid and lower thighs (10–15 reps each set)

Position yourself on a hack machine. Lower and raise yourself by bending and straightening your legs. Depending on the setup of the machine you use, you may find it advantageous to perform the hack squat with your feet placed in varying positions. (Heels together with toes pointing outward will develop the lateral section of the thigh.) Also, you may want to experiment by keeping your knees together or, alternatively, by holding them out to the sides.

Leg Extensions
Lower and middle thighs (10–15 reps each set)

Sit on a leg-extension machine with the tops of your feet (at the ankle-flexion point) affixed under the lift pad. Start raising the weight by extending both legs together. Do not "kick" the weight up. Start the lift slowly. If the machine you are using begins to race, you are exerting too much explosive force.

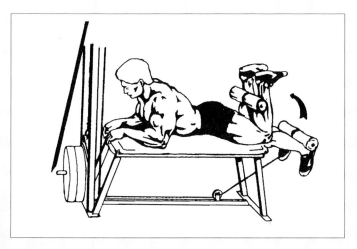

Leg Curls
Hamstrings (12–15 reps each set)
Lie on a leg-curl machine facedown. Hook your heels under the lift bar, and proceed to curl your legs upward in unison. Concentrate on securing the "feel" in the backs of your legs. Do not bounce the weight up after the legs straighten, but rather pause and start the curl slowly and deliberately.

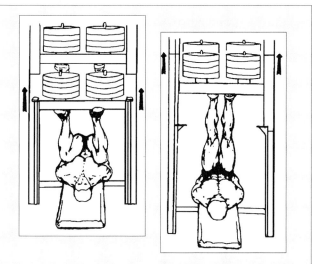

Leg Press
This is almost an inverted squat. Many bodybuilders prefer the leg press because it allows full concentration on the leg action without great involvement of the hip area. It is generally accepted, with good reason, that the leg press is not as effective a leg builder as the back squat, but it is certainly easier on the lungs.

Place your feet about 1 foot (30 cm) apart under the plate of the apparatus, and press upward until the legs straighten. Lower, and repeat. If you get stuck, aid the legs by using your hands to press on the thighs. Many gyms today have a 45-degree leg-press machine that is even more comfortable to use.

Sissy Squat
This is a very specialized movement, designed principally to work the lower-thigh area. Because of the unusual angle at which the exercise is performed, this movement is done with either no weight or only a moderate poundage. (Weight can be added either by holding a barbell in front of the shoulders or by holding dumbbells at arms' length at the sides.) The name "sissy squat" is not intended to denote that the exercise is easy, or for sissies. Quite the contrary. The name is derived from a character in Greek mythology, Sisyphus, whose eternal punishment by the gods consisted of having to roll a huge rock up a mountainside, only to have it roll down, over and over again.

The exercise is a little tricky. Adopt a position with your feet about 18 inches (45 cm) apart. Rise up on your toes, and lower into a squat while leaning as far back as possible. The point to bear in mind is to keep your thigh and torso in the same plane throughout the exercise. If performance is difficult, then hold on to the back of a chair or use a rope to prevent loss of balance.

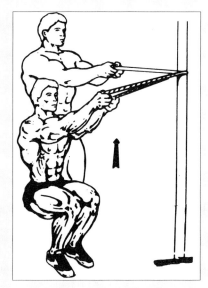

21

CALVES

BUILDING THE LOWER LEGS

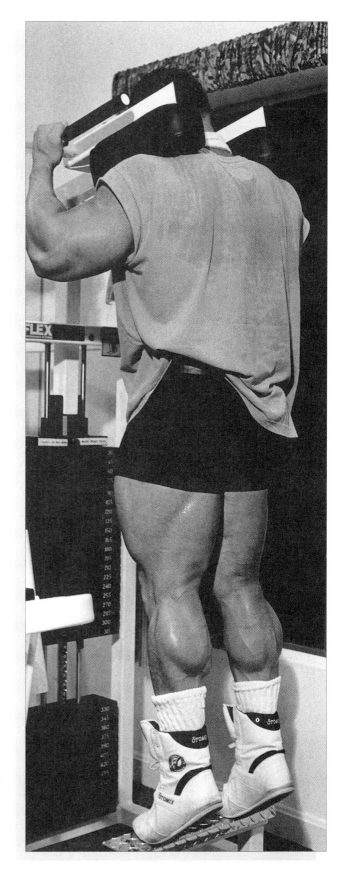

Hardly anyone apart from John Grimek had much calf development in the old days. It wasn't until the introduction and pervasive use of the standing calf machine, which enabled bodybuilders to use very heavy poundages each leg workout, that the lower legs really started to build. Today, we also have the seated calf machine and the leg-press apparatus, so it is no longer considered impossible to develop the calf muscle to any great extent.

Of course, if an individual doesn't have a sufficient number of muscle cells (genetic potential), he certainly will not have the ability to develop huge calves. Many men of African descent have a "high" calf, which in extreme cases means that they will not be able to build a balanced or big lower leg. Ironically, and in seeming contradiction to the above, some of the best calves in the business belong to black men. Witness the lower-leg development of Flex Wheeler, Paul Dillett, Aaron Baker, and Vince Taylor.

One of the first men to do a real job on the calves was the many-times Mr. Universe Reg Park. When he won the Mr. Britain, Reg had underdeveloped lower legs. But he totally remodeled them by working them on a daily basis with ever-increasing weight loads on a standing calf machine. At times, when he was unable to use the machine, Reg would perform donkey calf raises with two training partners sitting across his back. Just before the Mr. Universe contest, which he won, Park was training his calves twice a day. He ultimately put 4 inches on his calves with this heavy kind of progressive resistance training. How heavy? He used as much as 800 pounds of resistance on the standing calf raise.

Reg Park was the early hero of another physique star—none other than Arnold Schwarzenegger, the Austrian Oak. Curiously, Arnold also had difficulty in developing his lower legs. He decided to use Park's method of working the calf every day with extremely heavy weights. It worked so well that there was actually talk that Arnold had had silicone implants—a rumor with absolutely no foundation.

Way back in the nineteenth century when men wore tights, it was common practice for men to wear false calves, just as today we have jackets fitted with shoulder pads to give the appearance of added width. Eugene Sandow, known as the father of modern bodybuilding, actually did

(Right) Jay Cutler works the standing calf machine. (Left) Tom Varga displays a fine set of calves.

123

Chris Cormier works out on the seated calf machine.

wear false calves—until he developed a fine set of his own after a couple of years of training.

Another old-time bodybuilder who has really fine underpinnings is Boyer Coe. I watched carefully as he exercised them. Amazingly, Boyer spends at least 20 minutes each workout stretching his calves. He does this by standing on a high block without any weights, stretching up as high as he can, and then lowering down to maximize the effect. At this stage, Boyer is only interested in getting a full and complete stretch. Later on in the workout, Boyer trains his lower legs with resistance exercise (calf machines) and follows the traditional pattern of working them with about 5 sets of 15 to 20 reps.

Larry Scott, a pupil of Vince Gironda's and the first Mr. Olympia, says, "I prefer donkey calf raises, but when the calves start to burn really painfully, I bend my knees slightly to allow the pump to leave my lower legs and circulate more easily. This takes the pain away and allows me to continue on with blitzing my lower legs."

Way back in the late forties and early fifties, the bodybuilding world was in a state of shock over the superb calf development of Steve Reeves. No one could figure out how he got such shapely and well-developed lower legs. Writers and physical culture experts concluded that Reeves had built his calves primarily from having been a paper boy, cycling up and down the hills of Oakland, California. This was quoted as gospel truth for a decade, until Reeves exclaimed, in an interview published in *MuscleMag International*, "There were no hills on my paper route!"

In actual fact, Reeves worked his calves with a heel-raise machine and with donkey calf raises. He also redesigned his walk. He would rise up and down on his toes as he walked, so that his calves were put through an almost full range of extensions whenever he walked anywhere. His dynamic walking style still can be seen in some of his old *Hercules* movies, and it became a trademark of his virility and vigorous appearance. Ultimately, Reeves developed an even longer stride and a more pronounced arm swing. This became known as the "power walk," and a book by Reeves with the same name was later published. Needless to say, Reeves only uses this exaggerated walking style when exercising.

The point often is raised that those with high calves should not be penalized in a contest. Many black men, and not a few white ones, do not have the potential for building calves. Should they lose a contest just because they have poor calf genetics? Nature may have given us poor calf potential, poor arm shape, or lousy-looking abs. Too bad! The winner has to be the best-developed, best-proportioned guy. A man who is 5 foot 6 won't make a national basketball team, nor will a 6-foot-tall woman ever be an Olympic gymnast champion. By the same token, a man with 20-inch arms going steady with 14-inch underpins will not win a Mr. Olympia. Nor should he, whatever his genetics.

Who has the best calf development of all time? Although many champions have remarkable lower-leg development, I think that for all-around shape, definition, and size, the "World's Greatest-Looking Calves" title should go to the one and only Chris Dickerson. Roger Stewart, too, has amazing calves. Among the modern crop of bodybuilders with awesome lower legs are Dorian Yates, Flex Wheeler, and Vince Taylor.

These are three of the best lower-leg exercises.

Donkey Calf Raise

There is no doubt that the bent-over position one adopts for the donkey calf raise does something very special for the lower legs. This exercise is a real favorite of many bodybuilders. Lean on a bench or tabletop so that your upper body is comfortably supported parallel to the floor. Have a training partner sit on your lower back, over the hip area. Rise up and down on your toes until you cannot perform another rep. Use a 4-inch block under your toes to give greater range to the foot movement. You always should aim to perform at least 20 reps of this exercise. There also are a variety of resistance machines that can be used to duplicate this movement.

Standing Calf Raise

It is important that the calf machine you use be capable of loading on heavy weights. The apparatus should either carry a huge stack of weights or else be set up with a leverage benefit so that comparatively small weights provide a considerably increased overall load.

Rise up and down on your toes without excessive knee bending and without bouncing at the bottom of the movement.

Seated Calf Raise

This exercise, too, is performed on a special leverage machine. The principal muscle worked in this movement is the soleus rather than the gastrocnemius. Perform as many heel raises as you can, concentrating on maximizing total calf stretch with each repetition.

125

22
ARMS

FILLING OUT YOUR SLEEVES— BUT QUICK

To build arms, really build them, you need to train on a regular basis using about 10 to 15 sets for biceps and 10 to 15 sets for triceps—minimum! Some experts may disagree with this. So be it. I have no axe to grind. My recommendations are based solely on observation and experience. Perhaps one day there will be a concentrated training method by which we can activate the deepest muscle fibers of the arms with just one set of exercises. But until that time, be prepared to perform plenty of quality sets and reps.

When you compare the bodybuilders of the forties and fifties with the champions of today, you can see that arm development has advanced more than the development of other body parts.

We have come a long way from the Greek ideal, by which the neck, calf, and flexed upper arm were considered to be in perfect harmony if they measured

(Above) Lee Priest
(Left) The massive arm of Ronnie Coleman

the same. Until the middle of this century, most top bodybuilders aimed for this ideal and more or less achieved it. Steve Reeves achieved a calf, upper arm, and neck development of 18 inches, as did John Grimek, Clancy Ross, Roy Hilligen, Armand Tanny, Reg Park, and scores of others.

Today, we have title winners with 17-inch calves and necks, but 22-inch arms. Some have an even greater differential. The purists—and perhaps in my heart of hearts I am one, too—greatly disparage this new trend. No man should have upper arms that much bigger than his calves, they say. Right or wrong, the trend is with us. It is in vogue to have huge upper arms, and many bodybuilders feel, the bigger, the better.

It is a good idea to work the upper arms with at least one heavy exercise and one or two lighter pumping movements.

What follow are the best heavy, or "quality," movements for the biceps and triceps. It is recommended that you begin your biceps and triceps routines with one of these exercises.

127

Biceps

- Barbell curl
- Incline dumbbell curl
- Seated dumbbell curl

Triceps

- Close-grip EZ curl bar bench press
- Parallel-bar dips
- Lying triceps barbell stretch

With regard to biceps building, most champions today start their upper-arm routines with the two-handed barbell curl. Typically, they will knock off 5 sets of 10 to 12 reps to "hit" the biceps in the most basic manner. After this initial onslaught, two or three additional biceps movements are added, each of which will be performed for 3 to 4 sets of 10 to 12 reps.

When it comes to triceps building, there are literally hundreds of movements for this area—more than for any other body part. Still, a real size builder, without a doubt, is the parallel-bar dip. If you want big triceps, then this is the one to go for.

Avoid any exercise that gives you pain or discomfort in the elbow region. There are several triceps movements that can contribute to tendinitis (inflammation of the tendon) in the elbow. You must stop using such an exercise immediately or else cut down drastically on the weight you were using. Let's assume you have worked up to performing 6 reps with 120 pounds with lat-machine pressdowns, but the pain is unbearable. Either stop the exercise entirely or push it to the end

Alternate dumbbell curls performed by IFBB pro Ronnie Coleman

of your arm routine and merely perform a few pumpings sets of 20 reps with a far lighter weight.

Apart from proportion, it can be said that great arms have four essential qualities. Foremost is size. This includes thickness and roundness from the top of the arms, near the shoulders, to near the elbows. You also need good shape. This is largely hereditary. Ideal examples of champs with shapely biceps are Arnold Schwarzenegger, Don Long, Flex Wheeler, Lee Priest, Vince Taylor, and Ronnie Coleman. Men known for their triceps are Lee Priest, Kevin Levrone, and Nasser El Sonbaty. Veteran bodybuilders Robby Robinson and Mohamed Makkawy also have superb triceps development. Next, you need separation, the distinct delineation of the various muscles that make up the arms. The fourth essential quality is having vascularity and definition. The skin must be "thin," and you must have a low fat percentage. Veins grow in size along with the arms, especially if you achieve plenty of pump in your workouts.

A bodybuilder who starts his arm routine usually with barbell curls may switch suddenly to starting with chins, or even concentration curls. This is what bodybuilding is all about: shocking the muscles into growth. Lou Ferrigno, for example, changes his arm exercises every month. Even from workout to workout, he may change the angles if he feels the urge. He might, for instance, perform incline curls on a 45-degree bench one day but set it at a 30-degree angle the next. Alternatively, he may curl the dumbbells out a little farther from the body than usual. You, too, may want to alter your arm-training angles from time to time. You can, for example, try dumbbell curling with palms facing upward one day, and the next arm workout day do it with your palms facing inward.

Can you change the shape of your arms? Well, the answer is yes and no. You can add shape to your triceps by working, say, the outer triceps sections very hard. This will give an attractive appearance to the arms, especially when they are in the straight, or "hang," position. Alternatively, you can add impressiveness through exercises that work the lower triceps near the elbows. The biceps, on the other hand, are a little more stubborn. For example, by working the lower biceps on a shallow-angle Scott bench, you will "lengthen" the biceps only slightly, and by training hard on concentration peak-contraction curls, you will slightly increase the height of your biceps. But none of these changes will alter your inherited arm shape significantly.

Jay Cutler shows his form on the
EZ bar curl (start).

EZ bar curl (finish)

Preacher-bench curl (finish)

John Simmons trains his biceps on the
preacher bench (start).

Biceps Exercises

Incline Dumbbell Curl

Lie back on an incline bench slanted at about 45 degrees. Hold two dumbbells in the arms-down position. It doesn't matter whether you start the movement with your palms facing inward or upward. The only difference is that the forearms are brought more into play when the palms are facing inward.

Keep your head back on the bench, and curl up both dumbbells simultaneously. Your seat should not come up from the bench at any time during the curl, because that would aid the biceps in getting the weight up and therefore relieve them of some of their work. If they do less work, how can they build up size or strength?

As soon as the dumbbells reach shoulder level, lower, and repeat. Some bodybuilders actually tense their biceps at the end of the curl when the dumbbells are at shoulder level. This is just another way of maximizing intensity, and it is up to you whether you choose to do it.

Barbell Curl

This exercise has contributed to more 20-inch arms than any other movement. Hold the bar slightly wider than shoulders' width, and keep your elbows close to your body as you curl the weight upward until it is under your chin.

There are two distinct styles of doing this exercise: strictly (no leaning back during the movement, starting from a straight-arm position, using absolutely no body motion, or "swing") or cheating (hoisting the weight up by turning the trunk of your body into a pendulum on which the barbell can rely for added momentum). Both methods are workable, and most successful bodybuilders get best results by performing at least the first 6 or 8 repetitions in strict style and then finishing off the harder last 3 or 4 repetitions with a cheating motion.

Trainer Vince Gironda had his own way of performing barbell curls. He called it the body-drag curl. Basically, you hold the bar with a slightly wider grip than normal and you drag the bar upward along the body, instead of curling it outward away from the body. When you have lifted the bar as high as possible, you lower it and repeat.

Concentration Curl

Sit at the end of a bench. Exercise only one arm at a time, resting your elbow on the inside of the thigh (above the knee). Rest the nonexercising hand on your free leg. Curl the straight arm upward slowly, and then lower at the same speed. Concentrate intently on the biceps muscle as it contracts each time the dumbbell is curled upward. Immediately after training one arm, perform an identical number of repetitions with your other arm.

Arnold Schwarzenegger does another version of this curl, one he feels has contributed more to his monumental biceps size and peak than any other. He places one hand on a low stool or exercise bench and holds a dumbbell at arm's length hanging downward. Arnold makes a point of keeping his shoulder low throughout this exercise. This movement does in fact "hit" the biceps in an unusual fashion. It may take you a few workouts to get the hang of it, but once you do, I'm sure you will benefit enormously.

Alternate Dumbbell Curl

This exercise is a great favorite of champion bodybuilders. It works the biceps more directly than the two-handed dumbbell curl, because it prevents undue leaning back, or cheating.

Perform the movement by sitting erect and first curling one dumbbell. Then, as you lower it, curl the other arm. Lower slowly, and do not swing the bells up with any added body motion.

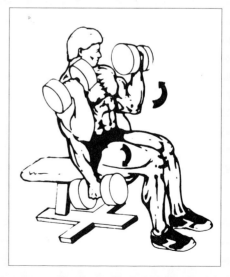

Scott Curl

Adopt a position with your arms over a preacher (Scott) bench. Hold either a barbell (as shown) or a pair of dumbbells. Curl up to the chin, and then lower slowly. Do not bounce the weights when the arms are in the straight position. Raise and repeat.

Standing Dumbbell Curl

Adopt a comfortable standing position with your feet about 18 inches (45 cm) apart. Holding dumbbells in both hands, curl both arms simultaneously until the dumbbells are next to your shoulders. Start with your palms facing inward. As you raise the weights, turn your wrists so that the palms are facing upward. Lower the bells slowly, and repeat.

Flat Bench Lying Dumbbell Curl

This is performed in the same way as the incline dumbbell curl, except that the bench is completely flat. Most people of average height or taller will need a comparatively high bench so that the dumbbells don't hit the floor at the bottom of the curl.

You may find that this exercise puts too much stress on your arms because of the unusual position. It is therefore advisable to start this exercise with comparatively light weights.

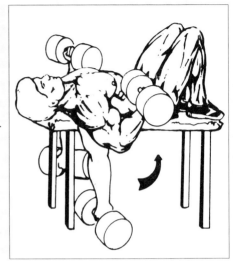

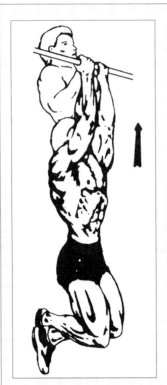

Undergrip Close-Handed Chin

This is a terrific biceps-building exercise, with a difference. Instead of the arm moving from the body, your body gets curled toward the arm.

Grasp an overhead horizontal bar with an undergrip so that your little fingers are 6 to 12 inches (15 to 30 cm) apart. Starting from a "dead hang" position with arms entirely straight, pull upward until your chin is above the level of the bar. Lower under control, and repeat.

Triceps Exercises

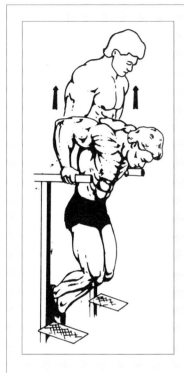

Parallel-Bar Dips

Mike Mentzer endorses this exercise. Start with your arms straight, feet tucked up under the torso. Lower (dip) while keeping elbows close to the body, and then raise and return. As you become strong enough to perform 10 to 12 repetitions, add weight either by "holding" a dumbbell between the thighs while crossing your legs at the ankle or by attaching iron discs to a "dipping belt" designed especially for the task.

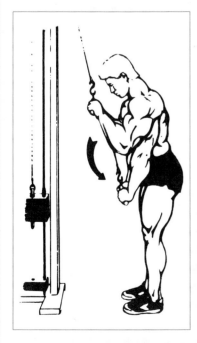

Pressdowns on Lat Machine

Was there ever a bodybuilder who didn't spend a great deal of time and effort performing this exercise? Start by holding a lat-machine bar with your hands 2 to 8 inches (5 to 20 cm) apart. Now press downward until the arms are straight. Return, and repeat. Most bodybuilders keep their elbows at their sides during this movement. A few hold the elbows out to the sides deliberately and "lean" into the exercise. The choice is yours.

Close-Grip Bench Press

This is a great favorite with veteran bodybuilder Larry Scott, the first Mr. Olympia, who said it has given him more triceps development than any other exercise. Lie faceup on a flat bench, feet firmly planted on the floor. Take a fairly heavy barbell (EZ curl bars are the most popular among the pros) from the racks (or have a partner hand it to you). Use a narrow grip so that your hands are only 2 to 3 inches (5 to 9 cm) apart. Keeping your elbows close to your body, lower the weight to your lower breastbone, and immediately push upward. If you are new to this or any other exercise, you should start by using only light weights, but many men do reps ultimately with 200 to 300 pounds.

Single-Arm Dumbbell Triceps Stretch

This triceps exercise develops the lower triceps area especially. With practice, it is possible to handle very heavy poundages—some top bodybuilders use dumbbells of 100 pounds or more. But it is not always advisable to use excessive weights in this particular movement, because they may put too much strain on your elbow joints and their surrounding ligaments. Beware of bouncing the weight out of the "low" part of the movement, as this, too, can cause elbow problems.

Bent-over Triceps Kickbacks

This is another favorite with today's pro muscle men. Bend over so that your torso is parallel to the floor. Hold a dumbbell in one hand, and hold on to a rail with the other hand. Raise and lower the dumbbell at an even rate, keeping the upper arm in line with your torso and parallel to the floor. Keep your upper arm tight against your waist throughout.

Seated Triceps Dumbbell Extensions

Hold a single dumbbell behind your back, with your upper arms as close to your ears as possible. Raise and lower the weight, while keeping your upper arms vertical. If the dumbbell you are using is adjustable, make sure the collars are secured tightly.

Lying Triceps Stretch

Lie on your back, as shown, and hold an EZ curl bar at arms' length. Lower it slowly to the forehead, and raise again to arms' length. This exercise works the entire triceps area, and it is considered an excellent mass-building movement.

Bent-over Lat-Machine Extensions

Assume a bent-over, strongly balanced position, holding a straight triceps bar attached to a lat-machine setup. Keeping the upper arms locked straight throughout the exercise, extend the forearms. Start the movement slowly, without jerking. This is an excellent exercise for the outer head of the triceps.

Robbie Robinson

23
FOREARMS

BRINGING THE LOWER ARMS UP

In the early days of the sport, top bodybuilders did no specific forearm exercises. Because every exercise (yes, even squats) works the lower arms in some way, it was argued that special forearm movements were not needed. Only Californian Chuck Sipes exercised his forearms individually with isolation movements. He was so different in his approach that even Arnold Schwarzenegger was heard to say, "Sipes spends too much time working his forearms." But times change, and now in the age of specialization, many bodybuilders find the time to train their forearms specifically.

Still, it must be said that many men have superb lower arms, yet have never trained them directly. They attribute their growth to "lucky genetics." The heavy-duty method of training (low sets, all-out intensity, forced reps) is said to aid forearm development even if the forearms are not worked directly with specialized movements.

Many bodybuilders complain that, as they complete a set, especially a heavy-duty set, their forearms "blow up" and fatigue earlier than the muscle they are supposed to be working. This can happen with exercises like chins, curls, and even upright rowing, and it's largely due to the individual's particular arrangement of muscle origins and insertions. Of course, it is a source of annoyance to those who experience it. Those of us with puny, underdeveloped forearms, however, would love them to blow up while we are doing sundry exercises, because our lower arms cannot be galvanized into significant development however we try!

Only a handful of bodybuilders have great forearms. Apart from Sipes, who is now no longer with us, old-timers Larry Scott, Tim Belknap, and Bill Pearl were noted for their massiveness in this area. Among today's champions, we see amazing forearms on Lee Priest and Paul Dillett.

There is not a great deal you can do with the natural shape of the forearms. Some men like Lee Priest have lower arms that, starting almost at the base of the hand, immediately sweep into rotund elegance. Others have "long" wrists, almost entirely devoid of muscle until well along near the elbow.

Because the forearms are in virtually constant use and have therefore developed a resistance to moderate exercise, they should be worked hard and with a sys-

(Left) Paul Dillett has some of the biggest arms in the world.
(Above right) Australia's Lee Priest

tem of high (10 to 20) repetitions. However, subjecting your forearms to progressive training is even more important than the repetition count. You will get nowhere by simply performing a few wrist or reverse curls at the end of your arm workout. You must attack your forearms with a planned campaign of ever-increasing workloads. Then you will reap the rewards of your disciplined endeavor.

Wrist Curl

Wrist curls work the flexors (the belly) of the forearm. Perform them in a seated position, with your lower arms resting on your knees (palms up) or on the top of a bench. Your hands must be free. Arnold Schwarzenegger keeps his elbows close, whereas other stars allow their elbows to be comfortably apart—anything from 12 to 18 inches (30 to 45 cm).

Moving only your wrist, curl the weight upward until your forearm is fully contracted. Allow the barbell to lower under control, and you may allow your fingers to "unroll" to some extent, but this is optional.

Reverse Wrist Curl

This exercise is performed in the same manner as the regular wrist curl, but your palms should face downward instead of upward. Also, you will be able to use less than half the weight in the reverse wrist curl. Most people find it more comfortable to keep the arms at least 12 inches (30 cm) apart in this variation.

Reverse Curl

Stand erect, holding a barbell at slightly more than shoulders' width. Allow the arms to hang down straight, elbows at your side, hands overgripped (knuckles up). As you curl the barbell, keep your wrists straight and level with your forearms, and keep your elbows tucked in. Then lower, and repeat. You will feel this exercise in the upper forearm, near the elbow.

Jay Cutler performs behind-the-back wrist curls (start).

Behind-the-back wrist curls (finish)

24

TANNING UP

NATURAL
AND
ARTIFICIAL
METHODS

It is every bodybuilder's dream to someday stand on stage looking huge, ripped, and tanned. Yes, a glorious golden tan is very important in achieving that look of super health and condition. How many times have you attended a competition where some men with fine physiques looked ultimately unimpressive because they lack a tan? We almost turned our head in disgust as our would-be champion flexed and squeezed to show off his larva-white physique. Without a golden tan, our friend had slim chances of taking home a trophy.

Many issues must be taken into account in our search for the ultimate tan. Sunlight could be viewed as the most important ingredient for tanning, but the alternative sources of coloring must be considered as well. For the competing bodybuilder, both tanning naturally from the sun and using blended artificial color make the ideal stage presentation.

In considering natural sunlight, you first must be warned that it can be very dangerous. The dire effects of overexposure cannot be stressed enough. In addition, there is absolutely no way that a tan can be rushed. The tan is nature's way of protecting you from ultraviolet radiation, and it can't be obtained in a single lengthy exposure.

(Above) Ericca Kern and Craig Titus
(Left) Lee Priest gets oiled up.

Your Skin

The skin consists of two layers that are separated by a thin membrane. The deeper layer, or dermis, is made up of blood and lymph vessels, fibrous tissue, sweat glands, hair follicles, and nerve endings. The outer layers, or epidermis, consist of basal cells that divide and form squamous cells. In a continual process, squamous cells die and produce the keratin layer, the outermost protective coating. The epidermis also contains melanocytes, cells that synthesize melanin pigment when exposed to ultraviolet radiation and produce the much-sought-after tan.

Tanning is the skin's response to ultraviolet injury and its attempt to protect itself from further damage. It is like drawing the drapes in order to protect your system from too much light, but the body needs time to react efficiently. Each time you expose yourself to ultraviolet radiation, it has to synthesize the melanin pigment that produces the tan.

Sunburn is only the initial damage caused by ultraviolet radiation. Prolonged exposure to ultraviolet rays also will interfere with the production of collagen fibers in the dermis, causing the skin to lose elasticity and creating premature wrinkles. Further deterioration of the skin's outer layer deprives the epidermis of nutrition and leads to atrophy of the skin, which is another name for aging. Finally, with increased injury over a number of years of sunbathing, the ultimate price of your tan could be skin cancer.

It is not my intention to discourage any body-builder from sunbathing to acquire a tan, but I must warn you to use caution. It serves no purpose to sit in the sun all day in the hope of acquiring a darker tan. Skin cancer can be avoided with a dose of common sense. You must appraise your skin type and determine what sort of complexion you have. Not all people can handle a lot of sun, and, yes, there are the unfortunate few who will never tan, no matter how long they expose themselves to the sun. As with everything else, genetics plays a big part. Fair people with blue eyes cannot hope to obtain as dark a tan as someone who has dark hair and eyes. Remember that a sunburn also can throw you off schedule, because the skin must heal before it can be exposed again.

Sensible Sunbathing

If you have a contest coming up, you should begin your exposure to the sun two or three weeks before the date of the contest. At the beginning, each exposure should be timed with care.

Before going into the sun, always apply a sunscreen to your lips and nose. I also would recommend that at this time you remove all body hair. This will give you an idea of how you are tanning. Certain areas sometimes can be very hard to judge if they are covered with lots of hair. Protect your eyesight by never looking directly into the sun. Ultraviolet radiation can injure the eyes severely, and even cause blindness.

Your first day of sun exposure should be no longer than 15 to 20 minutes front and back. I do not recommend using any lotion that will allow you to stay in the sun any longer than the prescribed time. Many of these lotions are advertised as containing sun-blocking agents that allow only the tanning rays to come through. But in my many years of tanning, I have never found this to be true, nor have I ever tanned any darker by using these lotions. I don't believe that any kind of lotion can improve your skin color—unless it contains a dye. Use a sunscreen on your body if you wish to spend extra time swimming or playing volleyball in the sun without burning. However, this additional, screened exposure to the sun's rays will not give you a deeper tan.

Like everything else, your rate of tanning is dependent on your genetics. Light-skinned people should be prepared to tan slowly. You simply cannot stay out in the hot sun for hours at a time. I also might mention that you can use a lotion or oil that has no sunscreen. This will provide a bit of welcome moisture for the skin, which can become very dry from the sun.

After the first day, study your body closely to see how your skin reacted to your first exposure. If you haven't burned, then increase your exposure 30 minutes per side on the second day. You will probably find that 30 minutes will cause you to redden a bit. It will take about three or four days at 30 minutes before you can increase the exposure. After four days, your skin probably will start to turn reddish-brown. The body's defenses have been mobilized into producing melanin pigment as a reaction to the sun's rays.

Be especially cautious during the final days of your first week of sunbathing. If you think 30 minutes is not enough and suddenly go for 45, the result could be detrimental in more ways than one. If you burn, you have severely injured your skin, and the healing process will take time. In addition, your first week's tan will peel off, so you will have to start all over again.

Spend your second week increasing your time in the sun by increments of 5 to 10 minutes, so that by the end of two weeks, your exposure will be 60 minutes per side. After two weeks, you should have a good base tan. The degree of your tan, of course, will depend on your complexion and genetics. In any event, with two weeks of exposure under your belt, you can now decide where you want to go from there.

It has been my experience that regardless of how long you now expose yourself to the sun, you will not improve on this tan. If a contest is near, then you might go for an hour and a half per side, but only for a week or so before the show. Once you have your tan, you can keep it golden by spending just an hour sunbathing two or three times a week. It serves no purpose to spend more time in the sun, and the dangers of hours upon hours of exposure have been spelled out earlier.

I remember a rather attractive young woman I used to notice at the beach. I was on vacation, and she must have been off for the summer, because I saw her spread flat on a beach towel every day. She just would lie there without moving, in obvious discomfort as temperatures climbed to the high eighties. At first, her body was a beautiful brown, but with increased exposure it started to look gray. Her exposure each day was five and six hours, and this must have gone on all summer, because after I had gone back to work, I still went to the beach on weekends, and sure enough, she was there. The last time I saw this poor soul came late that summer when I was shopping at a local mall. I walked into a clothing store, and there was the sun worshiper I had watched all summer. She did not have a healthy glow. She looked much older, and her skin was dry, flaccid, and dark gray. That's what the excess exposure to ultraviolet rays had done to her.

In summary, the sun is a very dangerous friend. You will accomplish nothing by spending hour after hour broiling in its intense rays. Even moderate exposure will leave you weak and listless, making it hard to train with your usual intensity. Unfortunately, competing without a beautiful tan likely will lessen your chances of walking away with a trophy.

Who Needs More Than Sunshine?

It may come as a surprise to natives of California and Florida, but 90 percent of the people in North America live in a climate where the sun is available to them only five to six months of the year. This leaves us northerners at an extreme disadvantage come contest time. In addition, a lot of our contests are in the spring, fall, or winter months. We are therefore left with the dilemma of finding a source of tanning other than a $2,000 trip to the south. Fortunately, there are artificial ways that allow the competing bodybuilder to obtain a less expensive tan.

Suntan Lamps and Tanning Booths

Some stores still sell what are known as "tanning lamps." However, a small tanning lamp is not going to give you an allover tan, because it is impossible to expose the entire body at one time. With these sunlamps, it's important to read the instructions carefully and follow them to the letter. Real damage can occur to the skin and to the eyes if you choose to ignore the rules of the lamp you buy.

Sunlamps can be useful for obtaining a little exposure before going on vacation and out in the sun. They allow the body to accustom itself to natural sunlight faster. Even though these relatively inexpensive lamps will not give you a glorious "Hawaii tan," they will give you some color, and that is better than nothing.

The expensive units can be found at the now very popular tanning salons. These artificial tanning booths are advertised as providing rays that tan without burning. However, if you burn easily in natural sunlight, then these lamps will burn you also. Thus, the value of these expensive salons will depend on your complexion. From discussing their merits with some of our pro bodybuilders, it seems that their only drawback is that the tan they give you will fade much sooner than a natural tan. Still, they can be of great benefit to a bodybuilder before a contest.

For anyone using these salons for a number of visits, the system of tanning and the exposure time will be determined by the host of the salon, who starts people off with a brief exposure and then increases the

Serge Nubret gets some sun.

exposures until a tan results. These visits may cover a period of months. The complaint often voiced is that in most cases everyone is treated the same. A person with a dark complexion can handle much more sun than someone with a fair complexion, so it would make more sense to tailor the exposure times to the individual's skin type.

Tanning Pills

In theory, when you swallow these pills, you tan from the inside out. They come under many brand names, one of the most popular being Orobronze. These pills contain a synthetic canthaxanthin, a coloring agent found naturally in certain crustaceans, fish, feathers, and vegetables. They cause the skin to take on a shade of orange-brown that varies according to the skin type. The pills are reportedly harmless and tend to work best in combination with natural sun.

The usual dose depends on your weight, surface area, and ability to assimilate this substance. It usually takes 15 days for an individual to show a color change, which peaks in 20 days. Don't expect too much. This sort of tan does not look natural, and it can leave you as orange as a carrot. But as a rule, your color will not change drastically, but just to a light orange, and in combination with natural sunlight, the pills can produce quite spectacular results.

One drawback of taking these tanning pills can be bright orange palms. For some reason, the palms of your hands seem to pick up more color than the rest of your body, and that can look rather amusing if not downright weird.

Bronzing Lotions

The most popular means of obtaining an artificial tan is probably the use of tanning lotions. Many such products are on the market. The results of these lotions depend on which brand you use, your skin type, and how well you apply them. I often have seen pro bodybuilders who had the stuff smeared all over them unevenly, so that they looked terrible and blotchy. Some had it mixed with oil, which made certain areas run more than others, and that gave them a patchy appearance with some parts of the body whiter than others. Avoid this sort of thing at all costs.

The secret of applying a good coating of a tanning lotion is patience. If you have a contest coming up on Saturday, apply your first coating late Thursday or early Friday morning. Your first coat should be light. Don't use a lot. Just dab a little on each body part, and then rub it in evenly. Rub very little on your knees and elbow joints, for they tend to absorb more than the rest of the body, and don't apply any to the face this first time. As soon as you have rubbed a light coating all over your body, including the back, just rest and let it dry. Do nothing that might cause you to sweat, or else the tanning lotion may run.

One point to remember is that your color will be much better if you can get some natural sun in addition. But even a sunlamp or tanning salon will improve the result.

With the first coat applied early in the morning, apply the second coating about midafternoon. Include the face this time, but otherwise follow the same procedure: very light, even applications to each body part. Under no circumstances should you try to apply a lot of lotion in an effort to obtain an immediate dark tan. It will take several applications. The final coat on Friday should be applied just before going to bed. Use the same methodical procedure. Dab a little on the arms, rub it in, then on the chest and stomach. Each area must be rubbed in completely before you go on to another.

The morning of the contest, you will find that your skin has taken on a fairly good color. The degree of the browning effect will vary from individual to individual, depending on the skin type and the product used. Some people will get quite brown, whereas others will tend to go orange. The day of the contest, you should try to apply two more coats: one early in the morning, and the second just before contest time. Any further coats will not give you a better tan. They will only give you a fake look and tend to shade your cuts.

One of the worst mistakes that many competing bodybuilders have made is to rub oil over the tanning lotion on their body before going on stage. Often the artificial tan will run with the oil and give the contestant a most unflattering appearance. Even some of our top pros have fallen prey to this, so be very careful about selecting a gloss you rub on in conjunction with a fake tan. Arnold Schwarzenegger was known to apply a light coat of Nivea cream before going on stage. Oil does give you better highlights, however. If you use it, make sure it is compatible with your artificial tanning lotion. Many contestants use Pam, the cooking-oil spray, with excellent results.

Conclusion

In achieving a good color for a competition, the most important factors are patience and moderation. Best results come from a tanning salon and/or natural sun in combination with a proven tanning cream.

By following the advice in this chapter, you can get yourself the best tan of your life, and that will certainly give you an edge in the competition. If you are not competing, your golden, even tan will make you look fantastic just the same. But remember, patience!

Points to Remember

• After sunbathing, always apply a good coating of skin-moisturizing cream to replace the natural oils lost during exposure.

• Your face may not be able to stand as much exposure as your body, so limit the direct exposure of your face.

• Always apply a screening agent to the extra-sensitive areas such as the lips, nose, and all around the edges of your swimming trunks.

• To obtain an even tan, remember to expose the underarm areas along with the rest of the body.

• Turn your towel or chair in the direction of the sun, so that the direct rays will fall on you evenly.

• Be aware that sand and water reflect ultra-violet light and thus increase its intensity.

• A hazy or cloudy day does not necessarily allow you a longer exposure time. The ultra-violet rays may filter through and burn you all the same.

• Be wary of sunstroke. Burning is not the only danger of spending too much time in the sun.

• You will become dehydrated from perspiration, so drink plenty of fluids.

• Do not stare directly into the sun, and close your eyes in a tanning booth.

• However flexible you are, you cannot apply a tanning cream efficiently to your back. Get help!

Jay Cutler

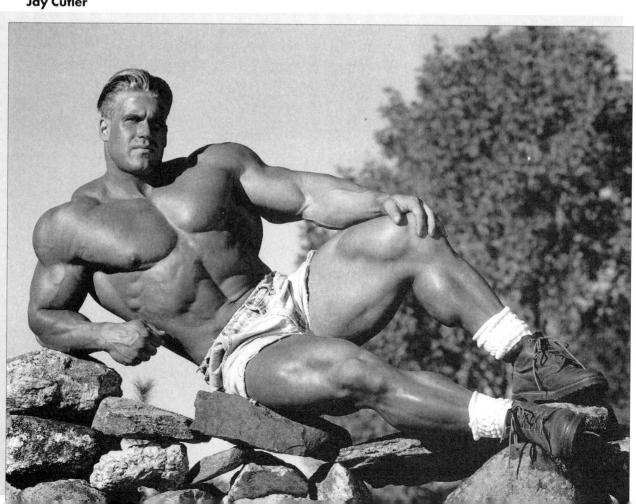

25

STATE-OF-THE-ART SUPPLEMENTS

AN OVERVIEW OF THE LATEST

Few topics in recent years have created as much controversy as food supplements. In one corner, there are those who maintain that proper eating habits will supply all the body's nutrient requirements. In the other corner are those who believe supplementing is an absolute must for hard-training athletes. And complicating matters further are outlandish claims made by supplement manufacturers themselves.

This all-new chapter will attempt to clear a path through some of the confusion, although it is not meant to be an all-encompassing guide to bodybuilding supplements. There are simply too many to cover in a single chapter. Instead, we will look at the new stuff—the products making all the headlines over the last few years.

Creatine

In their never-ending quest to achieve the ultimate in muscular development, bodybuilders and other athletes have been willing to inject, ingest, or inhale just about every concoction that comes on the scene. For the most part, their hard-earned money contributes to producing little more than the most expensive urine in town, but occasionally a supplement comes along that may in fact do what the advertisers claim: increase athletic performance. One such substance is creatine. Although it was isolated more than 160 years ago, creatine has fast become the hottest thing in bodybuilding.

Creatine is one of eight naturally occurring compounds synthesized from three amino acids: arginine, glycine, and methionine.

More than 95 percent of our total creatine reserves are located in our skeletal muscle, approximately one-third being free creatine and the remainder being in the phosphorylated form. Because the enzymes needed for creatine synthesis are located in the liver, pancreas, and kidneys, it means that creatine is produced outside the muscles and then carried to the muscles by way of the bloodstream.

The best natural sources of creatine are meat and fish. The average creatine intake from dietary sources is estimated to be one gram per day. As plants are very low in creatine, vegetarians rely solely on internal synthesis for creatine sources. Both forms of creatine are dependent on individual variation, and levels may be

(Left) Darin Lannaghan. (Right) An early shot of Reg Park, Hercules of the screen

influenced by such factors as age, disease, and muscle-fiber type.

To fully understand the importance of creatine, we first need a brief understanding of energy production in the human body. The primary energy source for skeletal muscle contraction is adenosine triphosphate (ATP). During exercise, ATP is broken down to form adenosine diphosphate (ADP). As long as energy is required, ATP must be replaced continuously, and once ATP is consumed, the body must dip into its creatine reserves. If creatine levels are low, fatigue sets in and exercise intensity is reduced.

As energy demands increase, creatine phosphorylate is broken down to produce free creatine and a liberated phosphate group. This high-energy phosphate group is then given, or "donated," to ADP to re-form ATP. The remaining free creatine is not wasted, because it is transformed to the phosphorylate form during periods of recovery.

You now can begin to see the importance of creatine supplementation. Although the debate rages as to whether creatine supplementation is beneficial for endurance athletes, the scientific and athletic communities agree that for sports requiring short bursts of strength and speed, creatine supplementation does boost performance. In addition, there is good evidence to suggest that creatine also may promote muscle growth.

How to Use Creatine Supplements

Although taking creatine in supplement form is a relatively new phenomenon in bodybuilding, studies conducted back in the early part of the twentieth century suggested that total creatine supplies can be increased by adding creatine-rich foods to the diet.

As supplements go, creatine is one of the easiest substances to take. It comes in the form of creatine monohydrate, a white powder that is tasteless and odorless. Although cold water can be used, you will find it dissolves more easily in warm water. Of course, at the other extreme, excessive heat will destroy much of the creatine.

For maximum absorption, take creatine on an empty stomach followed by a small simple-carbohydrate meal. The carbs will increase insulin release, which in turn increases creatine absorption. Some of the most popular creatine sources on the market include Twinlab's Creatine Fuel, AST Research's Creatine Complex-5, SportPharma's Creatine, and EAS's Phosphagen and Phosphagen HP. Perhaps the most powerful form of creatine monohydrate is MuscleTech's Creatine 6000ES. With each teaspoon providing 6 grams (6,000 milligrams, hence the name) of creatine monohydrate, it is probably the most potent creatine supplement available.

Other Sources

Besides creatine monohydrate, two other forms of creatine are available: creatine phosphate and creatine citrate. Although claims have been made that both are better than creatine monohydrate, creatine phosphate seems to be too unstable to be taken orally and there are no human studies of orally administered creatine citrate. It is possible that by the time this book goes to press, such data will be available, but as it stands now, stick with creatine monohydrate—it has a proven track record.

Loading

Manufacturers of creatine products recommend a loading phase of 30 grams a day for one week, followed by 5 to 10 grams a day for maintenance. Of course, these are average numbers, and larger bodybuilders might need 20 to 25 grams a day for maintenance.

Another important point is to spread the ingestion out over five to six small servings rather than one or two large servings. The reason is that, as with many supplements, the body can absorb only so much at a given time (vitamin C is well known for this). In terms of actual amount, this works out to about one teaspoon (approximately 5 grams) taken four times daily.

Or you might try one serving before and after working out. This increases creatine levels before exercise and speeds recovery afterward.

Although prices vary from state to state and country to country, in the United States a one-week loading supply (about 100 grams) of creatine will set you back about $25. Even by switching to the maintenance schedule mentioned earlier, regular creatine supplementation will cost $75 to $100 a month. In addition to your available finances, the level of your training and your bodybuilding goals also should be considered if you are thinking of adding creatine to your supplements.

Side Effects

As of yet, no studies have shown creatine to be toxic. Some individuals may retain water (something to keep in mind during the precontest season), but for the most part creatine has no known side effects. Having said that, a few words of caution are needed. Bodybuilders and other athletes no doubt will be taking creatine continuously for months if not years. Such a pattern of use has not been tested. In effect, today's bodybuilders are breaking new ground with each week of creatine use. The effects of such long-term high-dose use are unknown. While in theory there should be no problems, discretion is highly recommended. Because the body builds tolerance to foreign substances, it probably makes sense to cycle creatine usage. This also cuts down on cost.

Hydroxy Methylbutyrate (HMB)

HMB is another of the new generation of state-of-the-art bodybuilding supplements. Despite the limited scientific evidence available, the consensus among bodybuilders is that it holds great promise.

From a biochemical perspective, HMB is a metabolite (breakdown product) of the amino acid leucine. It also is produced by the body, although in small amounts.

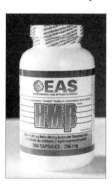

Studies carried out at Iowa State University found that subjects given 3 grams of HMB per day showed significant increases in strength and decreased levels of muscle damage following intense exercise. Although the exact process is not fully understood, it is believed that HMB decreases the effectiveness of enzymes that break down muscle protein. In this regard, HMB is not so much anabolic but, rather, anticatabolic.

In effect, HMB seems to counteract the protein breakdown often associated with intense training. Most bodybuilders who have used HMB report slow but steady strength and muscle gains. This seems to be the opposite of creatine, which tends to produce rapid weight gains and then tapers off.

Although an exact dosage has yet to be determined, the few studies carried out suggest 3 grams a day spread over three 1-gram servings. With regard to timing, there seems to be little difference between taking it on a full or empty stomach.

EAS introduced HMB to the bodybuilding world, but other supplement companies have recently jumped on the bandwagon, Twinlab being the most prominent. Most products come in capsule form, although EAS has gone a step further and produced a powdered version called Betagen, which contains HMB mixed with creatine.

Natural Testosterone Boosters

Dehydroepiandrosterone (DHEA)

DHEA is a steroid hormone produced by the adrenal glands. It is the most abundant steroid in the bloodstream, and it is present in even higher levels in brain tissues. DHEA levels fall with age, decreasing by 90 percent from age 20 to 90.

DHEA is a precursor of numerous hormones, including estrogen and testosterone. Bodybuilders take DHEA because there is evidence that some of it may be converted to testosterone and dihydrotestosterone. But this evidence is sketchy. Most studies conducted so far involved only older individuals with naturally declining DHEA levels. The effects on healthy young subjects are to all intents and purposes unknown. DHEA also has many other uses (all unproven); it is a potential antiaging, anticancer, and immune-strengthening substance.

With regard to dosages, I recommend 100 to 200 milligrams per day for men, and 5 to 25 milligrams per day for women. The women's dosages are lower because there is the potential for androgenic effects (such as a deepening of the voice) to occur.

It's important to note that as of 1997 the Canadian government has made DHEA an illegal substance. Although I question the merits of such legislation, possession of DHEA in Canada is, nevertheless, a criminal offence. So be warned!

Androstenedione

With the collapse of the Berlin wall, the world has suddenly become privy to the practices of former East German athletes. One such practice is taking androstenedione, which has fast become one of the most talked about supplements in bodybuilding.

Androstenedione is a metabolite that is made naturally by males and females and also found in the pollen of Scotch pine trees. Biochemically, it is the immediate precursor to testosterone. Of significance to bodybuilders is that, when taken in supplement form, some of it gets converted to testosterone by the liver.

The East Germans are believed to have been the first users of androstenedione. In the early eighties, East

German sports physicians developed an androstenedione nasal spray that, when administered, could boost testosterone levels for a couple of days afterward. They also experimented with androstenedione tablets and capsules but found that nasal sprays produced the best absorption.

Virtually all commercial forms of androstenedione supplements come in tablet or capsule form. This is why I haven't made my mind up yet as to whether androstenedione is all it's cracked up to be. While the research does show boosted testosterone levels, few if any studies have demonstrated increases in muscle mass among weight-training athletes.

As the research is limited, I won't nail down a precise dosage. Some bodybuilders report from experience that 50 to 100 milligrams taken twice a day produce the best results. Of course, other bodybuilders take many times that amount. But given the limited scientific data, I advise against this practice.

Although not an anabolic steroid, androstenedione may cause minor androgenic effects, such as acne, decreases in HDL cholesterol, scalp hair loss, and a deepening of the voice in women.

Tribulus Terrestris

This herb, another popular bodybuilding supplement, has been used for centuries as a homeopathic agent. Medicinally, it has been used as a diuretic, anti-inflammatory, antiseptic, and libido booster. It is this last point that attracted the attention of the bodybuilding community, as most libido boosters also may boost testosterone levels.

Unlike DHEA and androstenedione, which serve as precursors or building blocks for testosterone, tribulus is believed to boost testosterone by increasing levels of the luteinizing hormone (LH). LH is released by the pituitary when testosterone levels fall, much like the furnace cutting in when the temperature in your house decreases. Once released, LH stimulates the testes to increase testosterone production. Increased testosterone levels in turn tell the pituitary to decrease LH levels, and the cycle repeats. In theory, any substance that boosts LH levels also should increase testosterone levels. However, keep in mind that few studies have been conducted on tribulus to determine its effectiveness as an anabolic agent. The jury is still out on this one, as with androstenedione and DHEA.

It should be mentioned that the popular practice these days is to stack DHEA, androstenedione, and tribulus together. Although quantities vary, those used most frequently are 100 milligrams each of androstenedione and DHEA, and 1,000 milligrams of tribulus. The reports from bodybuilders look promising, but don't believe the hype of supplement manufacturers claiming that it works almost as well as steroids. This is just not the case.

Acetyl-L-Carnitine (ACL)

ACL is another of the new generation of natural supplements that may offer bodybuilders an alternative to anabolic steroids and other performance-enhancing drugs. Once again, I say "may," because the evidence for its effectiveness is still in the preliminary stages.

ACL is a modified form of the amino acid L-carnitine. Although the process is not fully understood, there is evidence to suggest that ACL may boost testosterone levels. Other studies suggest that it also may reduce cortisol levels and help prevent protein breakdown. In addition, researchers are experimenting with ACL as a means to slow the mental deterioration associated with Alzheimer's patients. Others are finding success using ACL to combat some of the side effects of diabetes.

The most popular supplement containing ACL at this time is MuscleTech's Acetabolan. Although MuscleTech's initial claims were met with skepticism by other supplement manufacturers, there now seems to be a race among them to get their own ACL-containing supplements on the market.

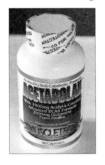

I'm not going to say that ACL is more effective than steroids or other drugs. But I will say that supplement researchers are on the right track when it comes to natural ergogenesis. It's only a matter of time before bodybuilding competitors using these natural supplements have physiques rivaling those of their drug-using counterparts of 15 or 20 years ago.

Insulin Boosters

Chromium

Perhaps the biggest revolution in nutrition concerns the attention given to single-substance supplements. And one of them, chromium, is fast becoming one of the most popular.

Most recognize chromium as the shiny metal on cars. Few, however, realize that it plays a major role in metabolic functioning. Chromium falls under the category of mineral trace element. Unlike minerals, trace elements are needed in only small amounts by the body. But don't be misled. Chromium and other trace elements are essential components of proper enzyme functioning. Being needed in small amounts doesn't diminish their importance.

The main function of chromium is to regulate blood-glucose levels. Although the process is not fully understood, it appears that chromium does this by increasing the binding power of insulin to cellular receptors. This means that, although it doesn't increase insulin levels, it does increase the effectiveness of any insulin available.

Perhaps chromium's biggest benefit to bodybuilders is its indirect role in amino acid uptake. Most know of insulin for its primary role in regulating sugar. But insulin also controls the rate at which amino acids are

absorbed into the bloodstream. In short, insulin is one of the body's most powerful anabolic hormones! Because chromium boosts the effectiveness of insulin, it only follows that it may increase protein synthesis as well.

Besides insulin regulation, chromium seems to play a role in reducing circulating cholesterol levels. Researchers at Mercy Hospital in San Diego found that healthy adults given 200 micrograms of chromium a day had their total cholesterol levels drop significantly. Although not conclusive, the results of this study and others suggest that chromium may offer those suffering heart disease an alternative to drugs or surgery.

The first chromium supplements came in the form of brewer's yeast tablets, made popular by the weightlifters of Muscle Beach back in the fifties and sixties. Although users swore to its benefits, brewer's yeast actually is a poor source of chromium—less than half of what is available in the biologically active form. Although other forms of chromium came on the scene in the sixties and seventies, it wasn't until the eighties that things improved. The first step in the right direction occurred when biochemists began to synthesize biologically active forms. They did this by adding inorganic chromium to live yeast cultures.

Beginning in the late eighties and early nineties, two forms of chromium supplements had come to

The mighty Arnold Schwarzenegger

dominate the market: chromium picolinate and chromium polynicotinate. Chromium picolinate is the complex form of chromium and picolinic acid, one of the metabolites of the amino acid tryptophan. Chromium picolinate has received its greatest praise from Dr. Gary Evans of Bemidji State University, Minnesota, who claimed that test subjects taking it had reduced body-fat levels, increased muscle mass, and lower cholesterol levels. On the surface, such results sound convincing, but few other researchers have been able to duplicate the findings of Dr. Evans' studies. The other popular form of chromium available, chromium polynicotinate, is chromium bound with niacin. But being niacin-based is not proof of bioactivity. Chromium can be bound to niacin in any number of different ways. And because no one is one hundred percent sure what the molecular arrangement looks like, synthesizing becomes a sort of supplement equivalent of Russian roulette.

As with most supplements, chromium should be taken with meals. This is because these substances act as modulators for the metabolism of other substances. For example, because insulin peaks following carbohydrate ingestion, and because chromium increases the effects of insulin, it only makes sense to take chromium when insulin levels are highest. Regarding dosage, it is suggested to take 400 micrograms per day. And as with most supplements, it probably makes more sense to take two 200-microgram dosages or four 100-microgram dosages a day rather than 400 micrograms in one lump sum.

Although rare, side effects associated with chromium use do surface occasionally. In many cases, the type of chromium being taken is at fault. Users of brewer's yeast (and those who work out next to such individuals) often experience severe discomfort in the form of flatulence. Although not life threatening, it makes for some noxious moments at the squat rack!

As a final comment, given chromium's effect on insulin regulation, those with diabetes or a history of diabetes in their family should consult their physician before using chromium supplements.

Vanadyl Sulphate

Vanadyl sulphate is a mineral form of the heavy metal vanadium. Vanadyl sulphate acts as a catalyst enabling diabetics to use the insulin in their bodies more efficiently. Vanadyl sulphate's importance to bodybuilders is that it triggers glucose transporters in the same way as insulin, causing increased glycogen levels and better assimilation of protein by muscle tissue.

The result is increased energy (from the increased glycogen levels) and decreased muscle protein breakdown (for use as an energy source). Together with a higher level of protein assimilation, the net results are faster gains in both size and strength. More energy means better workouts with heavier weights and more reps. I am particularly impressed with how pumped the muscles look when a bodybuilder is taking vanadyl sulphate (a result of the extra stores of glycogen).

Vanadyl is also an excellent supplement to stack with chromium and creatine. As both chromium and vanadyl improve insulin effectiveness and as creatine absorption is regulated by insulin, combining all three makes an ideal bodybuilding stack.

Everything has its flip side, however. This supplement may work too well. Because vanadyl sulphate has an insulinlike effect, thus regulating blood sugar, the pancreas (which contains the insulin-producing Islets of Langerhans) can become lazy. If the pancreas reduces the production of insulin, a Type II diabetic state can develop. Before using

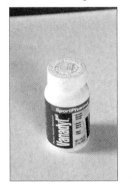

vanadyl, you might want to discuss this with your doctor.

There is a philosophical twist to this issue. Some of the experts I have spoken to are totally opposed to vanadyl sulphate, but will wax eloquent about chromium (which I also endorse as a daily supplement). The argument they pose is that chromium supplementing is fine because the dosages taken are equal to, if not slightly more than, the daily amount obtained from a well-balanced diet. But because vanadium is a trace metal, and the dosage supplemented is typically a million times more than the daily requirement, it is being used as a pharmaceutical. Perhaps the observation is a valid one, but as long as there's no harm being done, I really don't see a problem.

With regard to dosage, try taking 40 to 60 milligrams per day (30 to 40 milligrams for a body weight of 180 pounds.) I also would suggest cycling three weeks on, two weeks off. The body eventually builds a tolerance to vanadyl, as it does with most supplements.

Bigger than a house: Mr. Olympia Dorian Yates

Branched-Chain Amino Acids (BCAAs)

The three branched-chain amino acids are isoleucine, leucine, and valine. They are called BCAAs because they structurally branch off another chain of atoms instead of forming a line. Several studies have shown that BCAAs help to stimulate protein synthesis and inhibit its breakdown, so BCAAs have powerful anabolic and catabolic effects on the body. Some experts believe they also help in the release of some anabolic hormones, such as growth hormone.

In fact, the quality of a protein source can be based partly on its BCAA content. A protein source containing high amounts of BCAA is of high quality. Regular consumption of BCAAs helps to keep the body in a state of anabolism, or positive nitrogen balance. In this state, the body much more readily builds muscle and burns fat. In fact, studies of athletes taking supplemental BCAA have shown a loss of significantly more body fat than those not taking BCAAs.

So, how do BCAAs work? Within three hours of a meal, 50 to 90 percent of amino acid uptake into muscle tissue consists of BCAAs. This is because they are metabolized in the muscle. If an excess of BCAAs is consumed, the muscles are believed to absorb more amino acids to balance it out. To assist in this process, leucine stimulates the production of insulin. The result: blood glucose is taken up by muscle cells to be used as an energy source. The overall anabolic effect comes from insulin and the BCAAs working together to cause other amino acids to be used for building muscle tissue.

It's best to take BCAAs on an empty stomach with water just before you work out. A great product containing branched-chain aminos in quantity is MuscleTech's Acetabolan. Besides its main ingredient—Acetyl-L-Carnitine—it supplies 1,000 milligrams of leucine and 250 milligrams each of valine and isoleucine in one serving. Another excellent BCAA-rich supplement is MuscleTech's MESO-Tech. You get 12 grams of BCAAs per serving—that's the equivalent of 12 to 18 capsules!

Whey Protein

Up until the early nineties, the ultimate in protein supplements were those derived from whole milk and egg sources. Egg was thought to be the best because it had the highest protein efficiency ratio (PER), a scale devised by biochemists to measure how well a protein source is used by the body. Over the past few years, researchers have improved on egg proteins by making available a new supplement called whey protein. Often called "ion-exchange" proteins, whey sources have supposedly the highest biological value of any available protein source.

Whey protein is derived from milk serum and is the soluble part of milk protein. There are two purification techniques used to produce whey protein: ion exchange and ultra-filtration. Although both generate low levels of fat and lactose (a benefit to the large percentage of the population who are lactose-intolerant), the ion-exchange process is considered the better of the two techniques.

There are numerous reasons why whey sources are among the best forms of protein. They have higher amounts of branched-chain amino acids (BCAAs) than other protein sources. And BCAAs are not only used for building muscle tissue but also for oxidation during exercise. Another of whey's benefits is to elevate glutathione levels. Studies suggest that glutathione decreases may lead to Parkinson's and Alzheimer's diseases.

For bodybuilders, whey's greatest benefit seems to be in elevating insulinlike growth factor-1 (IGF-1) levels. Studies have shown that IGF-1 release is related to the quality of protein ingested, and whey protein seems to be superior to other protein sources in doing this. In addition, there is evidence to suggest that whey protein increases nitrogen retention, one of the conditions necessary for protein synthesis.

Rounding out the list of whey's benefits include its fast absorption by the body and a role in strengthening the immune system. From a practical sense, whey protein doesn't seem to result in painful bloating and gas problems caused by other protein products.

Despite the proclamations of some supplement manufacturers, whey protein does not have powerful pharmaceutical-type qualities. It's merely another great protein source that bodybuilders can use to their advantage.

Flaxseed Oil

This plant-seed oil is produced mechanically by a cold-pressed process. The oil contains alpha-linolenic (omega-3) fatty acid and can be used as an alternative to fish-oil supplements. Also found in flaxseed oil are linoleic (omega-6) and oleic (omega-9) acids. Flaxseed oil is believed to act as a thermogenic drug, causing the brown fat cells to burn extra calories as heat. At least that's the theory. Flaxseed oil received its biggest boost when the following colorful tale made its rounds throughout the gyms of North America.

Apparently, a couple of pros were known to be using a "secret weapon" that allowed them to get brutally cut before a contest. It became known that their magic bullet was flaxseed oil. Contest preparation involves forcing the body to give up its fat reserves. You can do this, but only up to a certain point, after which you can go on an almost fat-free diet and still be unable to get rid of the remaining fat. What these two pros did was go on an extremely low-fat diet; then, two weeks before the contest, they consumed huge amounts of flaxseed oil. The body, thinking it had huge amounts of fats in the diet, released its reserves. Of course, nothing is ever that simple, and keep in mind, these guys also were working out like mad. Further, as with most fats, essential oils can be stored as fat, if not burned as fuel.

Besides contest preparation, there's another reason to take flaxseed oil. Many diets low in fat may be nutritionally deficient. Most bodybuilding diets are high in protein, high in carbohydrate, and low in fat. Unfortunately, our body needs essential fatty acids for more efficient metabolism. Bodybuilders keeping their fat intake low year-round may be putting themselves at risk. Obese individuals with abnormal fat cravings (frozen yogurt, ice cream, and so forth) have been helped by being given supplements of flaxseed oil. The cravings were the result of a deficiency in essential fatty acids. Anyone who has dieted down for a contest will testify to the intensity of such cravings.

Please don't get the idea that this is only a precontest supplement. I personally believe that this supplement should be part of every bodybuilder's daily diet. I recommend 3 to 4 teaspoons for anyone less than 170 pounds, and 4 to 6 teaspoons for anyone more than that weight. Stacking flaxseed oil with fish oil and evening primrose oil will ensure that you obtain a rich and nutritional variety of essential fats in your diet, without adding fat to your waistline.

Besides flax oil, an excellent oil available to Canadian bodybuilders is seal oil. Doctors prescribe it for arthritic conditions, and researchers think it could be of great benefit to AIDS patients. The Chinese have used seal oil for more than 500 years as a traditional medicine. A company called J. Hiscock and Sons in Brigus, Newfoundland, manufactures the oil. American bodybuilders can't import seal oil because it is illegal to import products made from sea mammals.

Some have expressed concern over flaxseed oil because certain parts of the plant contain toxic chemicals. One compound, linamarin, can be converted by the body into cyanide. Now, this sounds fairly dramatic, until you consider that many foods contain such compounds. Flax oil is very safe. I haven't found one case of a person experiencing serious side effects. In controlled studies, people have consumed 60 grams daily without harm. When buying flaxseed oil, only purchase oil that has been cold-pressed, dated, and refrigerated.

Ephedrine

Ephedrine is a compound that is similar structurally to both adrenaline and the amphetamines, and it has many of the same benefits and problems. Ephedrine is found in plants of the genus Ephedra, the main commercial sources being Ephedra geraridiana (Pakistani ephedra), Ephedra nevadensis (Mormon tea), Ephedra trifurca (also called Mormon tea), and the famous Chinese herb Ma Huang (found in Dan Detainee's supplement, Ultimate Orange).

Ephedrine stimulates the beta adrenoreceptors, and, at low doses, it not only mimics adrenaline and amphetamines, but also improves mood and produces higher energy levels. Remember, I said "low" doses. At high doses, you can increase heart rate and blood pressure, perhaps leading to a stroke.

I get letters from bodybuilders all the time. Here's an excerpt from one with some questions about taking ephedrine.

Dear Bob,
I've been taking Ephedra capsules daily for over a year. When I first started, I could take one and I'd be pounding the iron for hours! Now I pop 5 before a

workout, and it may as well be decaf! What's going on? Should I switch brands? I've included the label off the bottle . . .

The pills were legit. There's a term in pharmacology called "tolerance." It means that after a specific time span, a drug dosage is no longer able to produce the same response. To get the same "kick," you have to go to a higher dosage. But you eventually reach the law of diminishing returns, which is what happened to our distressed letter-writer. Once the ephedrine starts losing its punch, cut it off. Taking a higher dosage will give you a bit of a kick for a few days, maybe a week, but then you'll be back to Square One. This stuff is safe if you use it right, but just increasing the dosage can lead to side effects you don't want.

If you are using ephedrine as a stimulant, I suggest following a four-week cycle. The first week, take one pill (25 milligrams) every second day. The second week, take one pill every day. The third and fourth weeks, take two pills a day, and then stop. Don't touch it for a month, maybe two. I know this sounds very conservative to many bodybuilders, but the beauty of stimulants is that you don't have to take a huge amount to get an effect.

The other mistake bodybuilders make is thinking they can skip sleep when they are on ephedrine. First, to make gains, the body needs rest to recover from your workouts. Second, a stimulant can work properly only if the body is well rested. Before you hit the pills, hit the pillow.

The Ephedrine/Caffeine/Aspirin Stack

Besides its stimulant effects, ephedrine also is used as a fat-burning agent. The combination of ephedrine, caffeine, and aspirin is probably the safest and most effective over-the-counter fat-burning stack available.

It is believed that ephedrine causes fat loss by thermogenesis, the liberation of fat stores by heat. By increasing the temperature of stored fat, the body can use it more easily as an energy source. Of course, to get this effect, you need to use higher dosages than if you are using ephedrine as a stimulant. The dosages used for fat loss (75 to 100 milligrams per day) quickly lead to tolerance. It comes down to goals.

What does the research say? In a study conducted at the University of Copenhagen, two groups of obese women were placed on a low-calorie low-fat diet. In addition, one group was given 20 milligrams of ephedrine with 200 milligrams of caffeine, three times a day.

The diet-only group lost 8.6 pounds of muscle and 9.9 pounds of fat. The ephedrine group lost only 2.4 pounds of muscle while losing 19.8 pounds of fat!

By combining ephedrine, caffeine, and aspirin (normally 300 milligrams), the body appears to produce more of the hormone norepinephrine, also known as noradrenaline. Being similar to adrenaline, noradrenaline stimulates the beta-3 receptors, which increases the rate of thermogenesis. There is also the possibility that ephedrine stimulates two thyroid hormones (T3 and T4), increasing the body's ability to burn fat.

If you decide to use ephedrine for fat loss, it might be a good idea to consult with your doctor first, especially if you have a history of heart disease or stroke in your family. After you get the clearance, try the following stack: 20 milligrams of ephedrine, 200 milligrams of caffeine, and 300 milligrams of aspirin. Take one dose a day for the first week, two doses a day for the second week, and then three doses a day for two weeks. Then go off all doses for at least a month.

Antioxidants

Simply put, free radicals are the disease, and antioxidants are the cure. When we breathe, we take in both stable oxygen and unstable oxygen. The unstable oxygen is what is known as free radicals, and they can have a harmful effect on cellular metabolism.

But free radicals don't just come from the air. They are found in the food we eat, and our bodies produce them as a result of daily activities. Intense exercise is believed to increase the formation of free radicals—it's probably the only negative aspect to exercise. However, before you give up exercising, keep in mind that exercise also boosts the body's ability to fight free-radical damage. It's sort of a trade-off.

Free radicals have been implicated in the development of chronic diseases, such as arthritis, and life-threatening illnesses, like cancer. A connection also has been made between free radicals and aging.

There is a way, however, to protect yourself. Consume antioxidants. The following are a few of the best antioxidants currently available.

Vitamins C and E

Vitamin C is a water-soluble nutrient. It is vital to anabolic processes and for the proper function of the immune system, and it has important biochemical relationships with many other nutrients. As an antioxidant, it protects the central nervous system (brain and spinal cord).

Vitamin E plays an essential role in the cellular respiration of all muscles. It increases cell stamina by enabling the muscles to function with less oxygen. This in itself decreases oxidation. Vitamin E also prevents both the adrenal and pituitary hormones from being oxidized.

Betaine

Although one of the least-known antioxidants, betaine, according to current research, may some day be considered one of the most important. Betaine is an oxidation product of choline. Betaine's ability to reduce homocysteine levels has made it one of the most promising heart drugs in recent times. High homocysteine levels are positively correlated with increased heart-attack and stroke risks. A 20 to 30 percent increase in homocysteine levels can cause a threefold increase in risk for heart attack and stroke. Bodybuilders take betaine supplements in order to enhance their digestive process, relieve indigestion, and as an antioxidant.

Most supplement manufacturers suggest one pill with each meal. But this is not recommended if you have peptic ulcers or already suffer from gastric hyperacidity. As always, check with your doctor and with your pharmacist before using it. If you develop any digestive problems, stop immediately and see your doctor.

Unlike most supplements, betaine is not obtained easily over the counter. In addition, to ensure purity, I suggest ordering from a legitimate supplier, like Sigma.

Glutathione

This substance is made from the amino acids cysteine, glycine, and glutamic acid. Glutathione's main influence is in the liver: increasing the organ's detoxifying abilities. It also plays a role in destroying free radicals that are produced from peroxides.

Tim Belknap

26
ROUTINES FOR ADDING SIZE

MAXING OUT THE MUSCLE

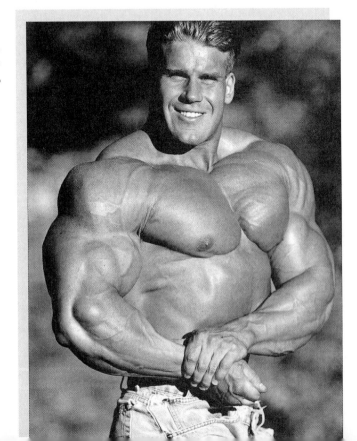

T he overriding question both beginners and advanced bodybuilders ask me involves the training routine. Everybody, it seems, wants to train with a routine that offers the maximum amount of return for the time and energy spent.

Training at Home

Let me deal first with the home trainer who has limited facilities with which to train. He probably has no lat machine, leg press, or leg-extension apparatus. Pek-Deks and the Scott curl bench are beyond his financial reach. But here are some training aids you simply must have, even if you need to make them yourself!

First, you should have a flat bench with weight support stands. It would be all the better if it is an adjustable flat/incline bench, but certainly a basic flat bench is essential. The only other absolute must is a pair of squat stands. These can be bought for less than $200, but many young people have been able to make their own squat stands out of wood, or got a friend or relation to make a set for them. The point to bear in mind is that without these important training aids, you can't really progress in either the squat or the bench press, both standard basic exercises. Needless to say, you also will need a 6-foot (110-cm) bar and a pair of 16-inch (40-cm) dumbbell rods, and enough weight discs to challenge yourself in the heaviest exercises. Ideally, of course, a set of dumbbells from 20 pounds to 100 pounds would be perfect, but few can afford such luxury.

Here is a very good, proven beginner's routine that will get you growing and keep you growing for some time. Do not make the mistake of thinking that more is necessarily better. *More people have gained more muscle on abbreviated weight programs than on any other system.*

Basic Routine

Warm-up	1-minute jumping rope
Press-behind-neck	3 × 8
Squat	3 × 10
Bench press	3 × 8
Bent-over rowing	3 × 10
Calf raise	3 × 25
Barbell curl	3 × 10
Triceps press	3 × 12
Sit-ups	3 × 20

**(Left) Jay Cutler shows an impressive chest.
(Above right) Don Long's awesome "most muscular" pose**

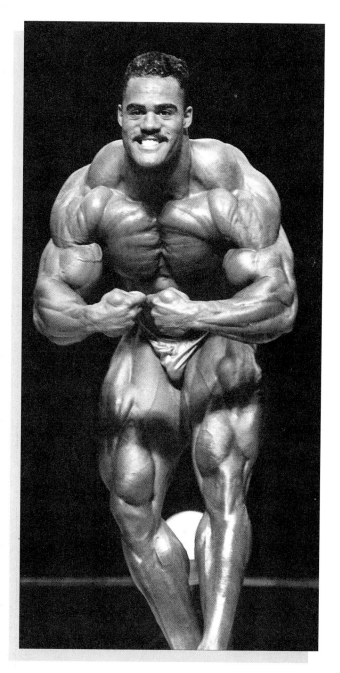

Incredibly, the above schedule can prove extremely effective for even the intermediate or advanced bodybuilder, at least for a period of three to six months. You will, of course, make better progress if you have a calf-raise machine for calf work in the above schedule. If you do not have access to one, then do the next best thing: hold a loaded barbell across the back of your shoulders for added resistance.

The next routine I'm going to give you is the super routine, for the more advanced trainer. This routine has been around for some time and is very popular.

159

The Super Routine

Warm-up	Jumping rope, 200–300 jumps

Shoulders

Seated press-behind-neck	4 X 8—10
Nonstop dumbbell raises:	
forward, side, and bent-over	3 X 10 each
Upright rowing	4 X 8—12

Quads

Squat (heels on 3-inch wood block)	5 X 8—12
Hack lift	3 X 10—12
Leg curls	3 X 10—15

Chest

Wide-grip bench press	
to upper sternum	6 X 6—12
Incline dumbbell bench press	
(45-degree angle)	4 X 10—12

Back

Prone hyperextensions	3 X 10—20
Wide-grip chin-behind-neck	4 X 10—15
Single dumbbell rowing	4 X 8—12

Calves

Heel raise (block under toes)	4 X 15—25
Donkey calf raise	4 X 15—30

For those who wish to split this routine, it is suggested that you divide it so as to train four days a week. You could, for example, divide it as follows, leaving the weekends free for family and relaxation.

Monday, Thursday

Shoulders
Chest
Neck and traps
Triceps

Tuesday, Friday

Thighs
Back
Calves
Forearms
Abs
Biceps

No Magic Schedules

There are many different ways to acquire a good physique, and not all of them involve the use of barbells and dumbbells. Wrestling, hand balancing, and gymnastics all add substantially to the musculature. It is undisputed, however, that the best methods involve the use of progressive resistance, but within this category no single routine can truthfully be called "the best." There is no magic schedule by which rapid gains can be secured or guaranteed.

By and large, bodybuilders are willing to do almost anything to accelerate their muscle growth. Thus, I would not hesitate to recommend any type of training madness if I thought it would help you get where you want to go. (Note, I said training madness, not steroid madness.)

I myself have done a heavy set of curls, 10 repetitions every half hour from 9 a.m. until 9 p.m., making a total of 24 all-out sets, in order to "shock" my biceps into growing. The result was a tumultuous headache the next day and very swollen biceps that appeared at first to have grown, but eventually diminished to their original 15˘ inches. I have trained the entire body every day for a month with little improvement. On another occasion, I did 50 sets of calf raises in three hours, and spent a week in bed!

I tell you all this in an effort to convince you that my thoughts are in no way conservative where bodybuilding is concerned. Because I know that you are looking for the best muscle-building techniques, I would not hesitate to recommend any exercise procedure to you if I thought it would work—even if the results were disproportionately small to the effort involved. But the truth is, there are no magic routines.

The Tough Old Days

Few people realize that on the whole the old-time bodybuilders trained far harder than the muscle men of today. Talking with Armand Tanny, a former Mr. USA and a very knowledgeable man on matters of muscle, he related how, in the golden days of Muscle Beach, it was quite common for him and his training partners to train all day long and think nothing of it.

He and his training buddies would do sets of repetition bench presses or curls that went on hour after hour from morning till dusk. One man, Zabo Koszewski, wouldn't even start his workout until he had done 1,000 sit-ups plus 1,000 leg raises! The 1951 Mr. America, Roy Hilligen, would do four hours of squats, as well as another four hours of presses, and then he would go on to finish his workout.

John Grimek, never a man to take the easy path, would do just about anything to get bigger. He has been observed performing scores of sets of just one exercise, and at one time he was doing repetition-free squats every workout. He performed thousands of nonstop repetitions!

The Deep-Knee-Bend System and Beyond

A more moderate routine that received a great deal of publicity in the muscle magazines back then and was considered rightly to be the superior technique at the time was the deep-knee-bend system for gaining muscular body weight. Looking at its development is worthwhile.

The value of this routine has to do with performing a few exercises for the large muscle groups. The system was mentioned first nearly 80 years ago in George F. Jowett's book *Key to Might and Muscle*. At that time, the average workout program consisted of at least 12 exercises, and sometimes 20 or more. It was not until 1930 or thereabouts that Mark H. Berry, editor of *Strength* magazine, began to appreciate the possibilities of a shorter type of routine with an enormous emphasis upon heavy leg work. Berry was greatly impressed by Henry Steinborn, the top-class pro wrestler and weightlifter of the era. Steinborn had accomplished an "impossible" 370 clean and jerk and had performed an incredible number of heavy squats in the course of his training at Sig Klein's barbell studio in New York.

When Mark Berry found out about Steinborn's training, he recommended similar training to other weightlifters for improving their leg strength and lifting ability. Rather to Berry's surprise, many students who took his advice began to report gains in body weight. This encouraged Berry to make further experiments.

Strength magazine had a wide distribution, so the deep-knee-bend method of increasing weight soon became a standard technique. Pupils all over the world tried it and made excellent gains. This led Berry to conclude that the deep knee bend (squat) and other heavy exercises were stimulating to the system. Then a method was devised that used no isolation exercises (unijoint movements). The basic routine that proved helpful to many barbell men of the day was the following:

Press-behind-neck
Squat (deep knee press)
Two-hands standing press
Stiff-legged deadlift
Bench press (press on back)

The three upper-body exercises were performed with one set each, 10–12 repetitions, and the deep knee bend and stiff-legged deadlift with 20 repetitions. The routine was to be performed three times a week, and no other exercise (such as abdominal work) was to be done. Emphasis was placed on the importance of adequate food and rest.

John Grimek, *the* physique star of the forties, reportedly used Berry's routine with great success. Ultimately, *Strength* magazine went out of business, but a young and vigorous publisher named Peary Rader was ready to promote the Berry system through his writings. Rader had started a typewritten, mimeographed bodybuilding bulletin and managed to gain a core of enthusiastic subscribers. The name of his bulletin, soon to become a respected, full-fledged magazine, was *Iron Man*.

With constant feedback from his readers, and in consultation with deep-knee-bend exponents Roger Eells and Jose Hise (who gained 29 pounds in one month with the deep-knee-bend system), Rader came up with a revised routine that omitted the stiff-legged deadlift but included the rowing motion, the straight-arm pullover, and the two-handed curl.

Peary Rader's Routine

Press-behind-neck	3 sets of reps
Squat	
(alternating with)	
Straight-arm-pullover	3 sets of 20 reps
Bench press	3 sets of 8 reps
Barbell curl	3 sets of 8 reps

Multi-Poundage

Following on from the Rader system, the British magazine *Vigor* came out with a useful program promoted under the name of the Atkin Multi-Poundage System. Like all other systems, it aimed to maximize gains in the shortest possible time, and it was practiced widely throughout the world.

The basic idea behind this multi-poundage system was to keep a muscle group working against maximum poundage throughout any number of repetitions.

In a normal system (as opposed to a multi-poundage system) a bodybuilder does not feel the weight until the last few repetitions. Atkin felt this to be a waste. Accordingly, his system (sometimes called triple-dropping) let a trainer start his set with a heavy weight. After a few repetitions, two discs would be removed. After another few repetitions, a further weight reduction was made.

When you first use this technique, it is a good idea to reduce the weight when you feel you could perform a couple more reps. Later, as you become more proficient, you can push until you cannot perform another rep. Incidentally, this method is definitely not for beginners.

When you apply this principle to dumbbell training, have one or two pairs of lighter dumbbells ready to use as you train. Then, when you can't go on pressing a pair of 50-pounders, lower them to the ground, and quickly pick up a pair of 40-pounders. For the last few repetitions, you can replace these with a pair of 30-pounders and train with those until you cannot do another rep.

It doesn't matter how you break up a set. Some people use a 3 ∞ 3 ∞ 3 system to make a set of 9 reps. Others work with 5 ∞ 5 ∞ 5 to make a set of 15 reps. Needless to say, this method, especially when using barbells, is not suited to the lone home trainer. You need assistance from two workout partners, one on either side of your exercise bar, to slip off a disc every few reps as you progress through your set.

Training in the 1950s

In the fifties Reg Park came to the fore in the world of physiques. Reg, at 6 feet 1.5 inches, was 240 pounds of bone and muscle and made every other bodybuilder in the world look small beside him, clothed or unclothed.

His training routine, too, was influenced by the deep-knee-bend phenomenon. Many thousands of weight men were affected in turn by Reg's training system, which was highly publicized in the *Reg Park Journal* he published as a monthly in Britain. The program below, which the many-times Mr. Universe used when going for maximum size, is a limited routine. Park, who trained with hundreds of different exercise variations in his career, is now more than 70 years of age and still considers it one of the best systems. Reg often split his routine in two and trained five or six days a week.

Reg Park's Routine

Prone hyperextensions	5 X 10
Press-behind-neck	5 X 8
Seated dumbbell press	5 X 8
Squat	5 X 8
Hack squats	5 X 10
Bench press	5 X 6
Flying exercise	5 X 10
Rowing	5 X 8
Chin-behind-neck	5 X 12
Barbell curl	5 X 8
Incline dumbbell curl	5 X 8
Lying triceps curl	5 X 10
Dumbbell triceps extensions	5 X 12

At around the same time that Park was training, Steve Reeves, whose good looks and super proportions appealed not only to bodybuilders but to the general public, was working out in a slightly different way. He trained the whole body three days a week, with a day's rest between each workout.

Actually, Steve trained very hard, with a Mentzer-like intensity, often performing negatives. For example, if while performing the incline dumbbell curls he could not get the weight up, he would assist with his leg by kicking the weight up, and then concentrate on lowering the dumbbell slowly. Because Reeves used such high intensity, he seldom found it necessary to perform more than 3 sets of any exercise. Steve liked variety in his training, so he would change his routine around frequently.

Steve Reeves' Routine

Upright rowing	3 X 9
Press-behind-neck	3 X 9
Lateral raise	3 X 10
Wide-grip bench press	3 X 9
Incline dumbbell press	3 X 9
Flying	3 X 12
Chin-behind-neck	3 X 11
Seated pulley rowing	3 X 11
Decline pullovers	3 X 11
Strict incline dumbbell curl	8 X 5
Triceps pressdowns	5 X 8
Single-arm lying dumbbell extensions	5 X 12
Barbell squat	3 X 6
Front squat	4 X 12
Leg curl	3 X 12
Calf raise	3 X 20

Longer Routines and Split Routines

Gradually, as bodybuilding contests became more popular throughout the sixties and seventies, routines became longer. It became common for serious bodybuilders to split their routines into two parts in order to shorten their training time.

Years ago, virtually no one performed exercises for the small areas of the body. Thigh biceps, serratus glutes, posterior delts, brachialis, rhomboids, erectors, teres, and tibialis were largely overlooked. Today, we pay attention to developing the smaller muscles of the body, so the schedules of the champs tend to be longer. However, as an aspiring bodybuilder, you should not try to emulate a champion bodybuilder's routine. Instead of gaining from it, you could easily tire yourself completely.

Here, for example, is a typical routine of Arnold Schwarzenegger, when he was winning the Mr. Olympia contest year after year. Bear in mind that Arnold frequently chopped and changed his exercises around, he said, "to keep my muscles guessing." Today, because Arnold does not compete anymore, he does much less. He is no longer motivated to build the huge 58-inch chest and 23-inch arms that brought him seven Mr. Olympia crowns and the title of World's Greatest Bodybuilder.

Arnold Schwarzenegger's Mr. Olympia Routine

Monday, Wednesday, and Friday Mornings

Chest

Bench press	5 X 8–10 reps
Flat bench flyes	5 X 8 reps
Incline press	6 X 8–10 reps
Parallel-bar dips (body weight)	5 sets
Cable crossovers	6 sets X 12 reps
Dumbbell pullover across bench	5 X 10 reps

Back

Wide-grip chins to front (body weight)	6 sets
T-bar rows	5 X 8 reps
Long cable pull	6 X 8 reps
Barbell rows	6 X 12 reps
High-rep deadlifts on box	6 X 15 reps
Single-arm dumbbell row	5 X 8 reps

Upper Legs

Back squats	6 X 10–12 reps
Leg extensions	6 X 15 reps
Leg press	6 X 8–10 reps
Leg curls	6 X 12 reps
Barbell lunges	5 X 15 reps

Afternoons

Calves

Heel raise on calf machine	10 X 10 reps
Seated heel raise	8 X 15 reps
Single-leg heel raise	6 X 12 reps

Forearms

Wrist roller curls	4 sets
Reverse barbell curls	4 X 8 reps
Wrist curls	4 X 10 reps

Tuesday, Thursday, and Saturday

Arms

Cheat barbell curl	6 X 8 reps
Seated dumbbell curl	6 X 6 reps
Concentration curl	6 X 10 reps
Close-grip press	6 X 8 reps
Triceps pressdowns	6 X 10 reps
Barbell French presses	6 X 8 reps
Single-arm triceps stretch	6 X 10 reps

Shoulders

Seated front press	6 X 8–10 reps
Standing lateral raise	6 X 10 reps
Dumbbell press	6 X 8 reps
Seated bent-over laterals	5 X 10 reps
Cable laterals	5 X 12 reps

Calves

Heel raise on calf machine	10 X 10 reps
Seated heel raise	8 X 15 reps
Single-leg calf raise	6 X 12 reps

Forearms

Wrist roller curl	4 sets
Reverse barbell curl	4 X 8 reps
Wrist curl	4 X 10 reps

After examining Schwarzenegger's schedule, you could conclude rightfully that he did just about everything. Each muscle area was worked, and worked hard. Unlike Dorian Yates, Arnold attributed his success to the concept of practicing "plenty of sets and reps."

Your Own Routine

Unless you are a very advanced bodybuilder, you won't be able to gain by using the schedule of Arnold Schwarzenegger. You will have to tailor his ideas and exercises to your own physical condition. Instead of five or six exercises per body part, do two or three. You can add exercises as your body gains in strength and condition. The important point to remember about a routine is that it must be balanced. You cannot grow if you are not working all the muscles.

Many beginners avoid heavy leg work like the plague. Even if your legs are naturally well developed, you should still work them regularly. The same holds true for all basic body parts. They should be worked on a regular basis. Of course, your weakest areas should be worked the hardest.

Another point to keep in mind is that a routine is simply a collection of workable exercises for increasing the size of your muscles, and not some magic formula. You must schedule exercises for each area and perform enough sets and reps to stimulate growth without inducing an overtrained condition. It's far better to adopt a short routine, and add to it as you progress in fitness and strength, than to tackle an overlong routine that will drive you to a sticking point. Start with a limited program, and add, add, add as you grow, grow, grow.

It wouldn't be right to discuss routines without mentioning the views of veteran bodybuilder Vince Gironda. (Vince died a week short of his eightieth birthday, in 1997.) It almost seems that Vince, the Iron Guru, was the originator of the routine as we know it today. His ideas were always revolutionary, and even 60 years ago he was ahead of his time.

Gironda cautioned that beginners should start off with only 3 sets of 8 reps (most authorities feel that one or 2 sets are enough for beginners). "After the first month," said Gironda, "I recommend that the beginner use 5 sets of 5 reps, the third month 6 sets of 6 reps, and, ultimately, after the sixth month, the trainer should be trying the advanced '8 sets of 8 reps' routine."

Gironda, of course, had experimented with numerous routines and their myriad variations. He concluded that the seasoned bodybuilder always can get an honest workout by performing a routine every other day consisting of doing one exercise per body part for 8 sets of 8 reps (except calves, which Vince said are "a high-rep muscle, and 20 reps minimum should be employed").

A typical Gironda workout schedule, which had proved itself successful time and time again at his North Hollywood gym (now closed after almost 50 years of service), is as follows:

Dumbbell lateral raises (lateral deltoid head)	8 X 8
Wide-grip parallel-bar dips (upper and outer pectoral area)	8 X 8
Seated lat pulley machine rowing (middle and lower lat area)	8 X 8
Kneeling facedown cradle bench triceps pulley extension (low, middle and outside biceps)	8 X 8
Body drag barbell curls (upper, middle, and lower biceps)	8 X 8
Front heels on block squat (middle and lower thighs)	8 X 8
Calf raise (gastrocnemius)	8 X 20
Crunches with weight (upper and lower abdominals)	8 X 8

The issue of tailoring your routine for your physical type has been mentioned earlier. Genetics determine whether you are a hard gainer or not. It doesn't take a bodybuilder long to discover that some men make exceptionally fast gains whereas others make only mediocre or even slow progress.

Although being a slow gainer can be discouraging, often there is some compensation in the form of "natural" shape and contour. Joints, for example, are often smaller (neater) in the hard-gaining enthusiast, and, accordingly, the muscles built around the wrists, elbows, ankles, and knees look more impressive when contrasted with these more aesthetic bony areas.

The hard gainer's approach to bodybuilding has to be better planned and more scientific and orderly. He cannot get away with partying, smoking, or boozing. His diet must be constantly at an optimum level, and it's suggested that he eat up to six to eight times a day (smaller quantities) to ensure that there is an ever-ready supply of nutrients for the body's needs.

If you are a hard gainer, then it will behoove you to limit the amount of exercises you do. Repetitions, too, should be curtailed, averaging around 6 or 8.

What follows is a recommended off-season training routine for the hard gainer.

Mondays and Thursdays

Shoulders
Press-behind-neck	4 × 6
Upright rowing	4 × 8

Chest
Bench press	6 × 6
Incline flyes	4 × 8

Trapezius
Barbell shrugs	4 × 5

Triceps
Parallel-bar dips	5 × 6
Lying triceps press	5 × 8

Abdominals
Crunches	3 × 15

Tuesdays and Fridays

Upper legs
Squat	5 × 6
Thigh curls	3 × 10

Upper back
Barbell rows	4 × 6
Chin-behind-neck	4 × 8

Forearms
Reverse curls	4 × 8

Lower legs
Standing calf raise	4 × 20

Biceps
Barbell curls	5 × 6
Incline dumbbell curls	5 × 6

If you are ever fortunate enough to visit Gold's or World's gyms in Santa Monica, watch the top bodybuilders train. Virtually every bodybuilder of note uses bench presses, flyes, squats, leg extensions, leg curls, presses-behind-neck, lateral raises, curls, triceps pressdowns, parallel-bar dips, calf raises, chins, T-bar rows, and crunches—all the usual movements.

Paul Dillett

165

Ronnie Coleman and Flex Wheeler

Dorian Yates' Routine

Multi Mr. Olympia Dorian Yates, of Sutton, Coldfield, United Kingdom, began bodybuilding in 1983. His first Mr. Olympia win was in 1992, and he has won it every year from then on. Throughout his career, Dorian has been tenaciously goal-focused. He has that single-minded determination that enables him to analyze scores of training theories, choose the best one for his individual needs, and tread fearlessly where no one had gone before.

Dorian has concluded that training routines for mass and strength should be brief and intense and followed by adequate periods of rest (72 hours or more) before working the same body part again. He has concerned himself with the basic exercises such as the squat, bench press, shoulder press, and rowing, realizing that they hit the major muscle groups that are capable of the most pronounced increases in mass. In the years leading up to visiting the United States, he had experimented with different routines, but decided that the split routine of dividing the body into two workouts and training four times a week is best. Following this method, he would train each body part twice a week, with adequate rest periods between workouts to ensure full recovery and accelerate muscular growth.

The following is the routine Dorian used to prepare himself to win the Night of Champions contest in 1991. Subsequently, he used this method of training to annex an unbroken string of Mr. Olympia titles.

Routine A

Chest
Bench presses	2 ∞ 6–8 reps
Incline presses	2 ∞ 6–8 reps

Back
Lat pulldowns	2 ∞ 6–8 reps
Barbell rows	2 ∞ 6–8 reps
Deadlifts	2 ∞ 6–8 reps

Delts
Presses-behind-neck	2 ∞ 6–8 reps
Side laterals	2 ∞ 8–10 reps

Routine B

Thighs
Squats	3 ∞ 8–10 reps
Leg presses	2 ∞ 10–12 reps
Leg curls	2 ∞ 10–12 reps

Calves
Calf raises	2 ∞ 10–12 reps

Biceps
Barbell curls	2 ∞ 6–8 reps
Incline curls	2 ∞ 6–8 reps

Triceps
Pushdowns	2 ∞ 8–10 reps
Lying extensions	2 ∞ 6–8 reps

Training Schedule
Day 1 — complete routine A
Day 2 — rest
Day 3 — complete routine B
Day 4 — rest
Day 5 — complete routine A
Repeat above cycle.

You should note that Dorian completed 2 or 3 progressive warm-up sets for each body part before attacking each exercise with 2 all-out sets performed slowly but under control, and without cheating on momentum. When he reached his failure, his training partner would assist him in 2 or 3 forced reps.

As an aspiring champion, you can use these same basic exercises, tailor them to your own stage of development, and one day you'll get there!

Michael Poulsen

27

TIPS TO BEAT OUT THE COMPETITION

THE WINNING EDGE

Everybody knows the names of the top bodybuilders: Flex Wheeler, Paul Dillett, Jay Cutler, Arnold Schwarzenegger, Gunther Schlierkamp. In fact, we are so familiar with these champs that we don't even have to hear their last names to know who they are.

But if you ever go to an amateur contest, especially at a national level, you will be awed by the number of really topflight physiques that are also around. Forty years ago, there were no more than a handful of men with 19-inch arms. Today, there are thousands upon thousands.

Even though there is an abundance of extremely well built guys around, there are not that many new muscle men breaking into the professional ranks. Of those who do, very few seem to be able to get their act together sufficiently to walk away with the top trophies.

It's a funny thing about bodybuilders—especially the big-name professionals—when they fail to win a contest, they often blame everyone else but themselves. If you have been to any seminars, you will know what I'm talking about. If a particular bodybuilder didn't win, he will tell you that the decision was political. If he placed way down the list, he will claim the judges overlooked him and didn't compare him with the top men. I have heard many different bodybuilders claim this one, yet in reality they were not compared because they did not merit it.

When the competitors come out on stage and go through the relaxed and mandatory poses, they are being judged, scrutinized, and compared right there. Lack of size, proportion, or muscularity is noticed quickly by an observant judge. A guy may look phenomenal in his bathroom mirror, but in the company of other top bodybuilders his great physique may pale beside a super-great one.

I remember when Arnold Schwarzenegger first came to the United States and entered the IFBB Pro Mr. Universe against Frank Zane in Miami, Florida. Frank beat him, and I wrote in *Iron Man* magazine that Arnold was undertanned and relatively smooth when compared to Zane. This was back in the late sixties.

A year after this contest, Chet Yorton held a party at his Malibu Beach home, and because I was in

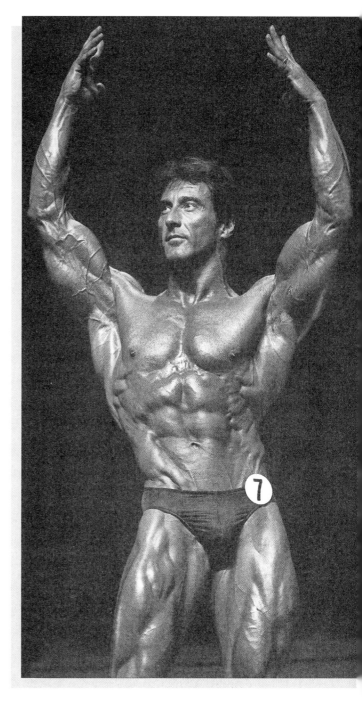

(Above right) Frank Zane is elegance personified. (Left) The ability to pose well is crucial in bodybuilding competition.

California at the time, he asked me to come. Almost as soon as the beer was flowing, Arnold started in on me for writing that he was smooth and pale at this show. Nothing I could say would calm him down. He insisted that he was in fact better than Zane, and that Joe Weider had arranged things so that he, Arnold, would have to pay his dues before winning the USA. He would not concede that Zane was better developed or better prepared.

Another year went by and I was again out with Arnold, this time at the Los Angeles Sporting Club, for dinner with Joe Weider. Had Arnold's ego relaxed sufficiently for him to concede defeat by Zane? No way! He even told Joe to his face that Joe had arranged the result because Joe "stood to make more money out of Zane winning." Joe just laughed at all this and told Arnold he was full of hot air. "You just weren't cut up enough, Arnold," retorted Joe between mouthfuls of steak.

Did Arnold ever concede that he was not in top shape that day in Miami when Zane trounced him? Yes, but it took him some 12 years. "Frank was better defined and tanned than I was," he finally coughed up. It took seven Olympia wins, innumerable successive victories over Zane, and 12 years' healing time before he had sufficient confidence to admit that maybe, just maybe, Frank had actually deserved that initial win.

This example is quite typical of the egos you will find in bodybuilding. As with acting, competitive bodybuilding demands that you "sell" yourself—one of the hardest things to do, especially on stage in front of a capacity audience. You simply cannot tell an actor that his acting is poor, and you can't tell a champion bodybuilder that he wasn't good enough to win.

If you suspect that your failure to win is due to "political" or other unfair reasons, then I urge you to consider that you may be wrong. There are, of course, poor decisions in bodybuilding. I have witnessed occasions when the wrong man won. But in the hundreds of contests I have witnessed, I have to say that for the most part the judges were genuinely sweating over making the right decision, and in the great majority of them the right man won.

Winning a contest is a matter of getting everything in line on the day of the competition. Size alone will never win a show, nor even size and definition, nor size and definition and proportion.

Everything has to be right, down to your posing trunks. The trunks you wear must be well fitting. The cut of the thigh should be high, and the color should blend, harmonize, or contrast effectively with the tone of your skin. Striped, banded, or multicolored patterned trunks are out. The judges are scoring you, not your pose trunks. You should keep them simple, so that they don't distract from your physique.

As discussed in an earlier chapter, you have to look tanned. Try to get the best natural tan possible, but even then, a few layers of Pro Tan or Jan Tana, or some other instant-tan concoction applied the last two days before a contest will help. Put on at least five layers (an hour or two apart). Very few natural tans are totally adequate, in that some areas (under the arms?) are less tanned than others.

Most men of African descent need to sunbathe and apply tanning creams to their bodies to look right on stage. There are a few who are so heavily pigmented that they could be considered too dark for the muscles to show up fully, but there are very few like this. I have known Americans, Egyptians, Puerto Ricans, Lebanese, Indians, even dark-skinned Italians, all competing without a heavy tan. They felt that they were dark enough. However, an Olympic contest will never be won by a man who is not sporting a maximum tan.

There have been several references in this book to the use of creams and oils during a contest. Arnold Schwarzenegger used to rub in Nivea cream two or three times and then replace his sweat suit and pump up. The resulting mixture of oil and sweat made his body glow with health on the posing plinth. Today, only a few bodybuilders use creams. The winners use a posing oil or Pam cooking oil.

Never completely relax on stage, even in the "relaxed" round. All eyes are on you, looking for flaws, so maintain an easy stride and a good posture. Hold your head high, and try not to keep staring down at your pecs or abs while on stage. Remember that charisma—the sparkling, confident personality that gives you power over the audience—is very important today, almost as important as muscles. It can certainly make the difference between winning and losing. Charisma is an elusive characteristic that few bodybuilders possess, but spectators recognize readily.

Make your poses dramatic, but not melodramatic. Good movement between poses (transitions) is essential, but avoid superfluous, exaggerated, or affected movements.

During the last month of preparation for a contest, practice your posing for two half-hour periods each day. The last week you should pose for two one-hour periods daily, and this includes "crunching" the pecs and flexing muscles in all positions, not merely in accepted poses. Practice keeping your thighs tensed for long periods. You will need this for the judging part of the show, because even when you are not being appraised directly, the judges will still be looking your way now and again, comparing you with whoever else is in the lineup.

It also will help to practice holding your poses for long periods (30 to 60 seconds). Compulsory poses can be improved distinctly by practice.

A week before the show, get a haircut and shave your body. Then shave your body again the night before the show.

If the contest is on a Saturday, your last precontest workout should be on Wednesday night or Thursday morning. The two or three days of rest will let you recuperate enough to be at your best at the contest. Do nothing else but pose during these last days. But no posing the morning of the show!

Sergio Oliva in his heyday would have a virtually complete workout to pump his body prior to competing. Today, much less pumping is done. Basically, stretching movements like wide-grip chins are performed, together with dips and curls for chest and arms. Overpumped muscles take on a bloat. Definition and separation are lessened when your muscles are gorged with blood. Some pumping is good, though. You feel stronger, bigger, and you are more alert mentally.

Today's top bodybuilders will do anything to stand out on stage. Witness the fact that Mohamed Makkawy made his stage entrance at one New York Grand Prix . . . on a camel! Ken Passariello made his face up like members of the rock music group Kiss and entered under the name of Demon. Eddie Robinson has made his entrance with a machine gun, and Aaron Baker has appeared in a Batman-like costume.

It is my opinion that posing will become more dramatic during the next few years. That is to say, some competitors will exhibit grace and poise, and will even dance during their free posing. I also think there will be a return to muscle control, as practiced by men of the past like Maxalding and Stan Baker of England, and Ed Jubinville, Bruce Randall, and John Grimek of the United States. To get the jump on this coming trend, I suggest you practice muscle-control techniques so that one day you will be able to wow an audience (and the judges) and bring home a first-place trophy.

When not posing during the last few days before an important show, try to get your feet up and rest. When you do arrive at the auditorium, you will invariably have to wait before pumping up. Hanging around can be a drain on you. Try to lie down somewhere, and do your best to relax. It takes the edema (bloat) out of the body, and that will give you one more advantage.

Flex Wheeler

171

Mohamed Makkawy

The last month's training before a contest is the most important of all. There is a fine line that you simply must travel. You should try to keep your exercise intensity high while limiting your food intake—no mean feat! The intensity factor—not necessarily more sets, but more effort with heavy weights—will help to keep your size. But fewer calories will mean less energy, and possibly slower recuperation. This can lead to overwork. If you find yourself shaking (hands trembling) during these precontest work-outs, you may be delving too deeply into reserve energy. This will not only stop progress, but will, in all likelihood, give your muscles a stringy, flat look. Even separation and definition can suffer when the depleted nerve reservoirs fight to normalize the body's condition. Because of all this, do not perform long workouts during your last four weeks of precontest training.

Bodybuilding is a thinking-man's sport, regardless of what the general public might feel. You need to pay

Vince Taylor

attention to every detail. You have to plan, experiment, evaluate, study, and analyze. However, my advice to you should be no more than food for thought. I do not know all the answers. You also should consider the advice of others with experience. Pay close attention to interviews with the champions of our sport. Interviews are the most factual type of writing to be found in the muscle press. Each magazine will inevitably contain gems of information that could make the difference between your winning or losing a competition.

The greatest experts, however, are not all title winners. Some have never even had an 18-inch arm. Among the most knowledgeable men in the field are Ken Wheeler, Leo Stern, Jack Neary, Arthur Jones, John Balik, Johnny Fitness, Greg Zulak, Will Brink, Steve Neece, Clarence Bass—not the biggest men in the world, but they know their bodybuilding!

173

28
POSING

THE ART OF SUCCESSFUL PHYSIQUE DISPLAY

W hat a man that "Buster" McShane was! I suppose you could say that he was larger than life. Buster was an Irishman, full of humor and zest. He was a multiple Mr. Ireland winner and an oft-time training partner of Reg Park, who rocked the world of muscles in the fifties and sixties.

Hardly had I left my native England in 1968 and settled in Canada, than Buster McShane paid me a visit. At the time, I was a teacher at a local high school, and Buster delighted all the students by giving them a weight-training demonstration. Buster and I had plenty of fun, training, drinking beer, and driving in my new Jaguar all over the Canadian countryside. After he left, my bench press quickly dropped the 30 pounds it had risen during his stay.

Buster was killed a few years later in Ireland, while driving home from a party in his new Jaguar V-12. Apparently, he had fallen asleep at the wheel and driven into a brick wall. Death was instantaneous. Those damn Jags always ran hot! Many a time I got drowsy on a long night's drive.

When I think of Buster McShane, I smile to myself because he had once had me laughing so hard that I thought I would split my sides. We had just stepped into an elevator on our way to the top floor of a big Toronto hotel. When the elevator doors closed, music piped in automatically. Lo and behold, it was "Legend of the Glass Mountain," Buster's favorite posing music! Without hesitation, he dropped to his knees and lost himself to the music, posing with heart and soul and complete abandon. Buster was drawing his routine to a close with a red-faced all-out "most muscular" when the elevator doors parted, revealing him to an audience of a dozen people. As we walked out of the elevator, convulsed with laughter, they gave us some pretty funny looks. Buster and I both knew that he simply had to pose when that music flooded the elevator. But none of the onlookers could be expected to understand his driving passion.

Posing shows you at your best or at your worst. Charles Gaines, of *Pumping Iron* fame, said it perfectly: "Posing is the art of the thing. Depending on how it is done, you can see in it either everything that is moving and beautiful and dignified about the display of a developed body, or everything that is ridiculous and embarrassing about it."

(Left) Flex Wheeler
(Right) John Sherman

The way you pose at a contest is the sum total of your endeavors. It's your showpiece. You can never achieve perfection, but you can reach out to grasp as much of it as humanly possible.

It's important to understand that in today's contests a bodybuilder is showing much more than his muscles, especially when it comes to professional competition. Along with having large, balanced, well-defined, and tanned muscles, our super-hero of the platform must exude confidence, have a feel for the shapes his body makes during and between poses, possess an air of mystical excitement and a flair for the dramatic, and show a decided penchant for communicating with the audience. In addition, today's competitive bodybuilder has to coordinate his posing to music. As a dyed-in-the-wool hard-core bodybuilder, you may resist this idea, but it is essential to ultimate success.

Some amateur contests do not allow competitors to select their own music, so here the problem of choosing the right music doesn't exist. Nonetheless, the aspiring bodybuilding should practice posing to suitable music. It will help him in the long run; even if he doesn't turn professional, he may be asked to guest-pose at shows or other functions, and his routine will look all the more polished if he has practiced the performance to inspiring music.

The question of what type of music you should pose to cannot be answered authoritatively, because styles change. In the sixties it was common for bodybuilders to pose to "Legend of the Glass Mountain." In fact, Reg Park wore out a half dozen records of this music, posing all over the world to it. Subsequently, "Exodus" and some Wagner became favorites, and then it was "Rocky" and "Eye of the Tiger." Currently, many bodybuilders are posing to hard-rock scores.

You must try to find the ideal piece of music for your routine. Your posing music must be, at least partly, known to the audience. It should accent your poses and build to a peak. In addition, it must draw the spectators into what you are doing on stage by touching either their emotions or their innate sense of rhythm. Many routines are tailored to the music "as is," but more and more today the music is edited and changed in content to suit the bodybuilder.

I have seen a contestant pose to a dull part of his music and get very little audience response, only to repeat the same poses to a more inspiring part of the tape and almost bring the audience to its feet. Music

has the power to make people cheer, dance, groan, or weep. In fact, it can motivate us to almost anything. Adolf Hitler roused a nation to accept his challenge, however evil and misguided, by his power of oratory combined with classical German music.

In *Muscle & Fitness*, musician David Lasker says that bodybuilders should use the musician's "trick" of counting during their posing: "Like a musician, you must count the beats in each bar of music. It will serve to pace your routine correctly." Actually, learning to count is easy. Your posing music is divided into measures, or bars, of four (or three) beats each. Count them: one, two, three, four. You have counted one complete bar, or measure. Some music, such as waltzes, has only three beats to a bar.

Lasker also suggests that you listen carefully for the strong and weak bars. A strong bar begins with a cymbal crash or a loud, accented chord. This is the moment to sweep in and hit a pose. During weak bars, you may simply hold your poses or glide into another attitude. Consider the number of poses in a routine (10, 20, 50?); then choose a piece of music that has the right number of strong bars to accommodate and do justice to your own posing display.

Many professional bodybuilders try to map out a routine by counting the strong and weak bars and then memorizing them in order to synchronize their poses with the chosen music. However, because little if any music was written with a bodybuilder's pose routine in mind, few pieces lend themselves completely to the bodybuilder's needs. Therefore, it is not uncommon for a pro bodybuilder to take his tape to a professional sound studio to have some of the music edited out in order to tailor the sounds to his needs. I can foresee the day when music will be written especially for the individual pose routines.

Show promoters allow contestants to bring along their own tapes, so if you are at all serious about bodybuilding competition, then you must practice until you can go through your routine almost automatically. Do not leave anything to chance. If you rely on the inspiration of the moment, then chances are you will fail to make the impression you wanted to make.

Whenever you practice your posing, count the bars (pose No. 1 lasts X bars, pose No. 2 lasts Y bars). This will help you duplicate your routine exactly, each time. It also will reduce on-stage "nerves," because you will have something to occupy your mind.

Unlike the compulsory poses—side chest, front abdominal, thigh, front and back double biceps, lat spread from front and back, and side triceps—the free poses are entirely up to you. It is important to select only those poses that highlight your physique. It's not a good idea to copy someone else's routine, pose for pose. Each of us is different, and by that virtue, your posing routine should be unique. You can get lots of good ideas for poses by looking at the various bodybuilding books and magazines.

When you practice posing, use a full-length mirror, and adjust the light to show your muscles to their best advantage. Try closing your eyes before you hit each pose. After you have fixed your position, open your eyes and check out the result.

I recommend strongly that you never give up on a pose. In the case of the compulsory poses, you simply cannot give up. They are your bread and butter when it comes to getting points. Obviously, when you perform before a judging panel, you will include only your best poses, but you should still practice a large variety. The flexing, stretching, and twisting will help you look better, and they will complement the work you have done in the gym.

If you find the art of posing particularly difficult, then by all means seek help from professionals who know about movement and stage presence. Ballet schools may be able to help. But before you enroll in a ballet class, make sure just what you are trying to accomplish is understood. It is not your intention to perform *Swan Lake* in ballet tights! Study the moves made by circus performers, Olympic floor gymnasts, and ice-skating champs. You can learn a great deal from them about movement and making shapes. Granted, it's not easy to strike up your poses before a panel of judges, or an audience, at least not as effectively as you do in front of your own bathroom mirror, but constant practice will make it easier and easier each successive time.

Although you can refer to books and magazines for individual poses, you will have to see live bodybuilders (if only on film or videotape) to learn the secrets of how to move between poses. The entire routine must be studied. It is a good idea to try to see a professional show, especially something as prestigious as the IFBB Mr. Olympia. Even so, some competitors will be far superior to others. Among the all-time great posers of the past are Sig Klein, John Grimek, Clancy Ross, Reg Park, Frank Zane, Ed Corney, Chris

Mauro Sarni of Italy

Dickerson, and Mohamed Makkawy. Among today's crop, we have Flex Wheeler, Shawn Ray, and Vince Taylor. I think one could say that Vince Taylor of Florida is the greatest living poser in bodybuilding today.

You seldom can learn how to pose from amateur contests (except at the national and international levels). Typically, each contestant starts with a double biceps and ends with a distorted, arm-flapping (to bring out the vascularity) "most muscular." In between, we are treated to an assortment of shuffles, groans, and grunts, accompanied by awkward attempts to flex up

Daren Charles, Aaron Baker, and Flex Wheeler

under duress. Occasionally, though, you will witness a posing phenomenon at an amateur show. The poser may not be the best built, and he may not even place in the contest, but his routine will be memorable. Watch and learn from these rarities.

Bear in mind that styles change. There are no rules "carved in stone" for the present-day poser. One could suggest glibly that a routine should incorporate drama, changes of pace and rhythm, smiles, and charisma. Yet who is to say that the next master poser to come along will make use of these features? If we could drop in on Eugene Sandow's first physique contest at London's Royal Albert Hall almost a hundred years ago, we would no doubt be bored to tears. Looking a hundred years into the future, it is just as likely that our descendants would be yawning at what we now consider a gut-bustin' display of muscle and might.

Basically, if you are lacking in size, you shouldn't spread out your legs or arms in your poses. Always remember that if you do stretch out, do so proudly. Nothing looks worse than a pointing arm that isn't

straight, because the poser feels insecure about fully stretching it out. Be proud: go to the extreme, straighten out to the limit. Have nothing, absolutely nothing, to be ashamed of, because you can bet your last dollar that it will be communicated to the audience. And never try to egg on the audience to applaud. They may respond, but they will resent being manipulated.

When you pose, your movements are being watched by every eye in the auditorium. A smile will let people know that you are communicating with them as you go through your routine, and a smile will beat out a frown any day. A good poser commands attention. The crowd will be eating out of your hands.

Unlike the work of exercising for muscles, in which overtraining will lead to staleness, the constant practice of posing, especially the performance of a complete routine, will serve only to fine-tune your presentation and improve your competitive edge. Master the art of physique display, and you are bound to be a winner. Fail to master it, and your chance of success is greatly diminished.

What a bod! Vince Taylor

29

THE 10 BODYBUILDING PITFALLS

SMALL LEAKS CAN SINK A BATTLESHIP

A

s with any other endeavor, bodybuilding has its pitfalls, and you must attempt to avoid them at all costs.

One of the earliest Mr. Americas, John Grimek, used to equate bodybuilding with a battleship. If it had a leak, or two or three, it would sink eventually. "You must seal your leaks," warned Grimek, and, of course, he was right. Overtraining, undernourishment, injury from poor exercise form, smoking and drinking—any bodybuilding pitfall—can be regarded as a serious leak. And any of them ultimately will affect your physical ability, mental resolve, and powers of recuperation.

Overtraining

It is difficult to overwork the muscles of a trained athlete, but easy indeed to drive the nervous system too hard. When a bodybuilder continues an exercise until he cannot perform another repetition, it is the nervous system rather than the muscle fiber that usually is unable to cope. According to physiologists, the first to fatigue in the neuromuscular system is the motor cells in the brain; next come the nerve "end plates," and then the muscle fibers. The nerves themselves are almost indefatigable. As a result of repeated muscular contractions, chemical changes occur at the nerve endings that make the transmission of the nerve impulse increasingly difficult, so that the brain has to provide a stronger stimulus via its motor cells in order to keep the repetitions going.

Forcing yourself in an exercise on a regular or prolonged basis past reasonable fatigue to exhaustion is therefore very expensive in nervous energy, and may even shut down your insulin production. The best bodybuilding gains are made by those who either limit their all-out training to infrequent intervals or drive themselves just far enough, but not too far.

Undernourishment

Are you, especially if you are a young beginning bodybuilder, getting adequate food for your muscles to grow? Often young people are walking "power stations." With their high metabolism, they burn up fuel at an unbelievable rate. To make matters worse, many young people miss breakfast. Without proper nourishment,

(Right) Gary Strydom
(Left) Lee Priest

181

how can you expect to have sufficient energy to not only supply your ordinary needs but also have enough fuel for adding solid muscle to your frame?

I am against pigging out to bring about body-weight gains, but you can't create something out of nothing. Maybe Rome wasn't built in a day, but let's face it, Rome would not have been built in even a trillion years without the bricks to do the job. Check over Chapter 13 (Ultimate Nutrition), and make sure you are feeding your muscles sufficiently for that all-important growth you want to achieve. Many successful muscle men eat five or six small meals a day.

Nutritional Overloading

In its own way, eating too much is as bad as not eating enough, and it is just as common. The unfortunate result is that for every inch you gain on your chest, you add two inches on your waist.

Once you gain weight, it's difficult to reverse the process—at least not without enormous dedication and perseverance. Once fat is comfortably sitting around your tummy, lower back, and hips, it is not dislodged easily.

Fat kills your appearance. Ironically, a covering of fat will make you look bigger in clothes (even though your pudgy face will give away your true condition), but when stripped at the beach or pool, you will look smaller than if you carried almost pure muscle.

What makes a muscular bodybuilder look good are the curves of his muscles—the way the delts run into the arms, the "peak" of the biceps, the delineation of the three heads of the triceps, the separation of the quadriceps—not to mention the absence of fat around the knee, ankle, and elbow joints. All these eye-catching features are lost if there is a preponderance of fat, for superfluous weight fills in the crevices between the muscles.

When your waist-chest differential diminishes instead of increasing, you know you are in trouble. Cut the calories before it is too late. Drastically reduce fat and sugar in your diet. You have become a resident of Fat City without even knowing it. Get out of town quick!

Faulty Proportion

Nothing inspires like success. When a bodybuilder notices that a particular body part—say, his chest or quads—is growing, he will undoubtedly work that area even harder. This, of course, leads to disproportion. The guy may have no arms or calves, or a lousy back, but his pecs and thighs are growing like crazy. Still in the wake of their progress, the lagging body parts look and become progressively worse.

Training for proportion is somewhat tricky because certain parts grow faster than others. The slower-growing parts probably have fewer muscle cells. Slow-growth areas must not be worked with the same number of sets or with the same intensity as the faster-growing regions—they must be worked with more!

Injury from Poor Exercise Form

Injury usually occurs from showing off, from attempting a new limit, or from sloppy exercise form. The most easily injured region appears to be the complex shoulder pectoral area, but any muscle can sustain a tear, strain, or other injury. Warm up adequately and use proper exercise style. If you bounce the weight up when doing Scott preacher curls, jerk the barbell when rowing, or "drop" into heavy full squats, you will only be inviting injury. And when it comes, sometimes it arrives with a vengeance and takes years to heal. Most injuries, however, can be avoided by working out wisely and using common sense.

Neglecting Aerobic Exercise

Many bodybuilders feel that aerobic exercise—prolonged exercise that works the heart and lungs steadily and increases cardiovascular fitness—is a waste of time. During a specific bulk or weight gain period, you may choose not to perform any type of exercise other than your heavy gym training, but aerobic exercise does have many benefits. It helps to keep your fat level down. It improves your wind during tough bodybuilding exercises like squats and so prevents your lungs from giving out before your legs. Aerobic exercise aids recuperation and keeps your appetite and bowels regular. It also hypes the metabolism.

Smoking and Drinking

In excess, both are killers. Drinking is tolerable in moderation; smoking is not.

I have known bodybuilders who smoked. Although some had their trophy-winning days, none lasted the course. The simple fact is that a bodybuilder needs an abundance of energy and pep. Cigarette smok-

Kevin Levrone and Paul Dillett pose-off in the heat of competition.

Darin Lannaghan

ing diminishes both. What suffers first is the squat. The cigarette smoker just can't do the reps. Soon his workouts deteriorate to becoming a joke, and ultimately his physique mirrors his low-quality training. If you smoke cigarettes, your days of hope for bodybuilding success are numbered.

You can drink in moderation, especially with meals, but not when you are preparing for a contest. Arnold Schwarzenegger always liked a glass or two of wine or champagne, but during the days leading up to a contest not a drop passed his lips. If you like to down a bottle of whiskey or fill your stomach to the limit with beer over the weekend, then you are overindulging. Your drinking has got the better of you, and success will elude you as far as contest-winning bodybuilding is concerned. Moderation or abstinence is the key.

Extra Exercise

Do not bemoan your inability to gain weight if you in any way resemble the proverbial human dynamo who fritters away his energy in a weekly routine of soccer,

bowling, two evenings of racquetball, three evenings of roller-blading, and dancing over the weekend. By all means enjoy occasional friendly games and pastimes, but when you are trying to add muscle to your frame, keep other physical activities in perspective. Be prepared to abandon other physically demanding pursuits until you have obtained and consolidated your required muscular gains.

Relying on Steroids and Other Drugs

Anabolic steroids are not a substitute for hard work, good food, and scientific supplementation. However, they will increase the liquid retention of your muscles, which will make them bigger in appearance. The buffer effect also will allow you to lift heavier weights, but too many young people use steroids instead of tough training. Therefore, they are often in poor physical shape. The only pose they can present effectively is the "most muscular." Anything else makes them look "bunched up."

I am against the use of steroids, but even if I had no second thoughts about their safety, I still would not recommend their use. They won't make you look any better. For one thing, they tend to bloat and thicken the waistline.

Other drugs that bodybuilders use, including diuretics, all cause physical damage of one kind or another. Those who rely on them for muscle and shape never come out on top in the end. As a bodybuilder, you may feel invincible. But if you neglect your health, you will quickly find out that you are not.

Listening to So-Called Friends

Sometimes a friend, or someone who pretends to be, will tell you what you want to hear rather than the truth. Many a time I have overheard "friends" telling compet-

ing bodybuilders that they "wuz robbed," when in reality they were lucky to place where they did. It is hard for a bodybuilder to know how he looks up on stage in comparison with other bodybuilders, so he relies on what others tell him. What it all comes down to is being honest with yourself.

I am not suggesting that you should adopt a negative attitude toward yourself, but nothing is worse than acting as though you should have won a contest when the entire audience knows differently. In one show I attended, two competitors who failed to place in the Top 3, threw their 5th- and 6th-place trophies at the judges sitting in the front row of the audience. On that occasion, the winner was deservedly in first place, and those two aggressive maniacs were lucky, in my opinion, to have done as well as they did.

Craig Titus shows extreme vascularity.

30
QUESTIONS AND ANSWERS

HELP!

Lateral Triceps

Question: I want to develop the lateral (outer) head of my triceps. I greatly admire this area of the arm, but only a handful of bodybuilders seem to have exceptional development in this region.

Answer: When fully developed, the lateral head of the triceps does add drama and quality to the way the upper arms look. In fact, the entire arm benefits. Most two-handed triceps exercises in which the hands are closer together than the elbows work the outer head vigorously. Some pros use the kneeling overhead pulley press, whereas others use the close-grip bench press with EZ curl bar and the triceps lat-machine pressdowns with the elbows held out to the sides.

The single-arm lying dumbbell stretch effects great stress on the lateral triceps head. The dumbbell (start with a light weight, because perfect style is essential) is lowered across the chest, so that when the right arm is exercised, the dumbbell is lowered to the left pectoral.

How Much Muscle?

Question: I have just begun bodybuilding. How much muscle can I expect to gain in the first year? How do I know what potential I have?

Answer: No one can say for sure what your potential is. It's hard to assess by measuring, weighing, or looking at you. Many champions were skinny to begin with. Beginners, however, often gain quickly. If you eat well, train progressively on a regular basis, and sleep well, you will be on the right track for gaining 10, 15, 20 pounds or more during your first year. Beyond this, body weight can be gained by forced feeding, but a great deal of what you gain will be fat. Intermediate bodybuilders seldom gain more then 7 or 8 pounds of pure muscle a year, and often much less. Steroid abusers gain considerably more, but they risk their health, big time.

Cellulite

Question: I have been exercising with weights as well as stretching and running for more years than I can recall, but lately I have noticed a dimpled effect on my backside. I need your advice on how to rid myself of this unsightly appearance.

(Left) Ronnie Coleman

Answer: What you have on your seat is fat! It has become fashionable to refer to this dimples effect as "cellulite," but in fact it is simply fat. It tends to be stored in large amounts on the upper thighs and bottom and to dimple the skin because it is pulled downward by gravity.

The beauty industry has come up with numerous so-called cellulite cures, such as creams and gels to massage in, pills to break up fat and prevent fluid retention, heat treatments to draw out toxic body fluids, and injections of "anticellulite" serums. Although some of these methods may prove marginally successful, it's best to redesign your diet so that your overall calorie count is lowered and you greatly reduce your fat and sugar intake. In addition, you may want to increase the speed and severity of your workouts so that more calories are burned. By attacking the problem on two fronts, you are much more likely to have success. You will not lose those dimples easily without changing your diet. All junk food is out of the question.

Bigger Wrists

Question: How can I enlarge the size of my wrists?

Answer: As you grow older, up to somewhere in your midtwenties, wrist size increases naturally. All weight trainers, especially those who train with heavy weights, gain some wrist size due to actual bone growth (thickness) and to development of the sinews and tendons that surround the wrist. There is no specific growing exercise for the wrist, but it does thicken and strengthen with regular weight training.

What Are They Doing?

Question: Whenever I turn the pages of a bodybuilding magazine, I am amazed at how fantastic and "ripped up" the muscles on the various bodybuilders look. How do these men do it? I train extremely hard, eat well, take protein supplements, and have developed some pretty good muscles, but I can't even imagine myself ever looking as defined and trained as those men in the magazines. What are they doing that I don't do?

Answer: First, let me set you straight. The top bodybuilding champs are not always in such super-ripped shape. In fact, often their super muscularity is evident only for a few days around the time of a contest. The pictures you see in the magazines are nearly always taken within a couple of days of an important competition. Many, in fact, are taken on the day of the com-

187

petition, either while they are posing on stage or else under studio lights set up backstage. So, remember, when you see top bodybuilders "ripped to the bone" in the various magazines, they are not like this all the time.

I also must tell you that many bodybuilders peaking for a contest resort to potentially dangerous drugs, like pharmacy-grade diuretics, coupled with a near-starvation diet. Therefore, on the day of the show they are prone to cramps, nervousness, and even heart irregularities, which can cause lasting health problems and may shorten their career as well as their life span.

After the peak has been reached, there is a natural and necessary period of relaxation. Some champion bodybuilders actually give up all training after a contest and follow some other interest or activity. However, most bodybuilders, especially the hungry ones who want to get to the top in a hurry, work out all year round. As a hard-training bodybuilder, you have a choice of either looking big and beefy all year round or looking slightly smaller but still muscularly impressive. What you cannot be is big and super ripped all year. That is possible only at the conclusion of a planned peaking period, which requires tremendous motivation, a restricted diet, and intensive training.

Veins

Question: I want big veins in my chest, back, and legs. I will be entering a contest soon, and I need to develop extreme vascularity. Should I take large doses of niacin and thyroid?

Answer: Definitely not! Taking thyroid can knock out your own thyroid gland, and high-dosage niacin can knock your eyeballs to the back of your head. Forget about them!

Why on earth would you want maximum vascularity? Ugly, bulging, crisscross veins usually run contrary to the lines of the muscle separation and detract totally from any shape or mass you might possess. Your thinking is way off base. Veins don't make a physique, muscles do. A degree of vascularity is fine, but too much is the trademark of an amateur, not of a winner. If you had your way, the judges would be eyeballing your ugly veins and nothing else.

Headaches

Question: I have been getting a lot of headaches lately. They seem to come when I squat or do heavy pressing exercises. Often they go away when I finish my work-

out. Could my training be causing these headaches? I enjoy using forced reps and negatives and would hate to have to ease up on my training.

Answer: Sometimes training increases internal pressure, and this could cause your headaches, but there are a thousand and one causes of headaches. The room could be too hot, you may have high blood pressure, it could be the stress factor or excessive noise—the list is endless. I have no solution other than to tell you that I have had headaches from training, and so have many others. They often have occurred when I went a bit crazier than usual with my workout, especially after a layoff. Perhaps you are not quite ready for forced reps yet. I suggest you reduce the intensity and length of your training for a while and attempt to build up slowly again. If the headaches persist, see your doctor.

Forearms Blow Up

Question: I hope you can help me with a problem I have had since Day One of my training. Whenever I do biceps and triceps work, especially biceps, my forearms blow up like crazy. The pump hurts so much, I have to stop a set before my upper arms are fully worked. My forearms are 15° and my upper arms are only 16° fully flexed. I feel and look like Popeye. Please tell me which upper-arm movements I can do so that my forearms won't blow up each set.

Answer: Many men have this same problem, and in one sense you can consider yourself fortunate. You may never have to do specialized forearm work because your forearms will always receive adequate stimulation from regular upper-arm exercises. Many bodybuilders have to do 10 to 12 sets of forearm movements just to keep their lower arms "in line" with their upper arms.

Your particular muscle-cell allocation and leverage makeup will dictate which exercises you can do without getting congested forearms. You will have to experiment with the different barbell, dumbbell, and pulley exercises, including EZ curl bars, to find out which exercises you can do without provoking your forearm problem.

Overwork

Question: I love bodybuilding, have been doing it for years, but every time I complete a heavy workout, I get either a sore throat or a cold, or a case of the runs (diarrhea). I desperately want to be a champ. Do you have any advice for me?

Answer: You are overworking. You need to backtrack a

little. Make your workouts less exhausting. Then gradually build up the intensity again. Do it slowly. Do not miss several workouts and then suddenly go all-out. This would bring back your problems. Overwork puts the system in shock and is counterproductive to successful bodybuilding. To maximize your gains, make sure you are taking a top-of-the-line supplement, such as *MuscleMag's* Formula One elite series.

Sets

Question: I would like to know why some bodybuilders perform 20 or more sets per body part and still exercise with a great deal of intensity, while others use the same high intensity but only in a couple of sets.

Answer: Twenty sets are more demanding than a couple of sets. In fact, 20 sets are more demanding than 19. There are, however, a lot of other variables, such as the amount of time you have available for training, your body's ability to recuperate, the tolerance of your metabolism and nervous system for strenuous and prolonged exercise—the list goes on. One thing is certain: doing more sets is entirely useless if it involves overtraining and not fully recuperating between workouts. This is why the Mentzer System is so popular. It is based on the fact that you only need a minimal number of sets (1–2) to zap the muscles into growth.

What is needed for regular gains is progression. Intensity is of prime importance, but not everyone can keep on increasing intensity, workout after workout. The next and probably easiest step is to increase the number of sets. The problem here is that the more sets you do, the more the muscle cells are stimulated, but it is very much a matter of diminishing returns. In other words, 20 sets per body part are better than 10 sets, but only fractionally. You have to perform a great many forced sets to effect smaller and smaller benefits—provided you recuperate between workouts. Without full recovery, you will drive yourself, not to the top of bodybuilding, but into the ground. Most bodybuilders perform 3 to 6 sets of each exercise.

A Joule?

Question: I am naturally heavy, so am constantly on a diet and buy just about every book on the subject. Recently, I came upon a new word, which I just don't understand. Could you please tell me what a joule is?

Answer: Joules (J) are going to be used to measure energy. One calorie is equal to 4.2 joules, and 100 joules is a kilojoule. One million joules will be known as a megajoule. For the time being, you may as well stick to calories—everyone else probably will.

Stretch Marks

Question: I am only 26 years old, but I'm worried because I am getting stretch marks at the sides of my chest, and even on my arms. It is so bad that I may even quit training. What do you suggest I do to get rid of them?

Answer: Some people think stretch marks are caused by growing too fast: your muscles are growing faster than your skin. In fact, faulty nutrition should take much of the blame. If you do not feed your body the essential nutrients it needs, then it will rob nutrients from the least essential organs, often the skin. This makes the skin less elastic and more likely to tear under strain. To help avoid stretch marks, you should supplement your body with a good B-complex supplement, vitamins A, D, and E, and pantothenic acid. For stretch marks that have already occurred, you can try to minimize the damage by rubbing them with vitamin-E oil (the d-alpha tocopherol type). Also, stretch marks can be removed surgically, but the results are not always satisfactory.

Belts

Question: Do I need a lifting belt? I see that many bodybuilders wear these in their training, but I do not want to waste my money if a belt is not really necessary.

Answer: Most bodybuilding, as with weightlifting, involves overload training. Heavy training can put a stress on the lower back. It is definitely advisable to use a belt in all overhead lifts, rowing, deadlifts, and all forms of squats. A belt diffuses the strain of heavy poundage and so reduces the chance of injury to the lower back. A belt is like an additional set of waistline muscles. Look at the seasoned squatter who never uses a belt: his tummy is way out to here!

A belt will not only keep you tight for your exercises and help to ward off injury, but it also will add to your strength once you get accustomed to it. Movements like the barbell or dumbbell press, deadlifts, and squats will improve 10 percent or more. Remember that a belt doesn't have to be drawn tight all the time. Most trainers wear it loose and only tighten it up (very tight) just before an important set of heavy exercises. Immediately after completing that set, they loosen it again. Belts are not needed for bench-press or abdominal exercises.

INDEX

A

abbreviated training routine, 56, 159
abdomen, exercises for the, 102–107
acetyl-l-carnitine (ACL), 150
"active rest," 42
aerobic exercise, 65
Alexeev, Vasili, 63
alternate dumbbell curl, 128, 131
alternate dumbbell front raise, 87, 92
alternate dumbbell press, 88, 89
anabolic process, 67, 69
anaerobic exercise, 65
androstenedione, 149–150
anterior head, 87
Anthony, Melvin, *33*
antioxidants, 156–157
Apperson, Lee, *73*
arms, exercises for the, 126–135
Arnold Press, 90
Arnold Schwarzenegger's Mr. Olympia routine, 163
Atkin Multi-Poundage System, 161–162
Aykutlu, Hamdullah, 17, *44, 79, 109*

B

back, exercises for the, 108–115
back press, 87, 91
back squat, 117
back stretch, 39
Baker, Aaron, *21, 29, 61,* 110, 123, 171, *178*
Balik, John, 11, 173
Bannout, Samir, 103
barbell curl, 128, 130
barbell-row movement, 26
barbell squat, 117
basic routine, 159
Bass, Clarence, 41, *63,* 64–65, 71, 173
Belknap, Tim, 137, *157*
belts, 189
benches, 27–28
bench press, 87, 88, 95, 96, 97, 99
Benfatto, Francis, 103
bent-over flying movement, 87, 88, 92
bent-over lat-machine extensions, 135
bent-over rowing, 38, 88, 112
bent-over triceps kickbacks, 134
bent-torso pulls, 39
Berry, Mark, 161
betaine, 157
biceps, exercises for the, 127–128, 129, 130–132
body-fat percentage, 62–65
body types, 14–19
Borg, Bjorn, 10
branched-chain amino acids, 154
Brink, Will, 173
bronzing lotions, 142, 144
bulking up, 75

C

cable crossovers, 101
calcium, 73–74
calf raises, 125
calorie-dense foods, 72, 73
calves, exercises for the, 122–125
"carbing up," 52
carbohydrate, 71, 72
Cardillo, John, 41
catabolic process, 67
cellulite, 187
Chaney, Dr. Warren, 25
Charles, Daren, *119, 178*
cheat reps, 31–32, 52
chest, exercises for the, 94–101
chromium, 150–156
close-grip bench press, 27, 133
close-handed floor dips, 52
Coe, Boyer, *10,* 42, *43,* 46, 92, 124
Coleman, Ronnie, *24, 40,* 109, *110, 119, 126, 128, 166, 186*
Columbu, Franco, *36,* 37, 45, *58,* 59–60
competition, beating the, 168–173
compound training reps, 35
concentration, 46
concentration curl, 131
Connors, Jimmy, 10
Cormier, Chris, *31, 53, 98, 124*
Corney, Ed, 177
Cottrell, Porter, *8, 93*
creams and oils, 170
creatine, 147–149
Cureton, Jr. Thomas, 63
Cutler, Jay, *25,* 49, *115, 123, 129, 139, 145, 158,* 169
cycling, 40–43, 81

D

definition, 63–65
dehydroepiandrosterone (DHEA), 149
delts, 87–93
Dente, Garard, *86*
Dickerson, Chris, 124, 177
Dillett, Paul, *13,* 17, 49, *66,* 68, *71, 93,* 102, 103, *116, 136,* 137, *165,* 169, *183*
dips, 96, 128
disc-loading barbell, 28
Dobbins, Bill, 52
donkey calf raises, 55, 123, 125
Dorian Yates' routine, 166
drugs, 60–61, 64, 184–185
dumbbell flyes, 96
dumbbell pressing, 87, 92, 96, 99

E

ectomorph, 11, 15–19
Eells, Roger, 161
endomorph, 11, 15–19
enzyme levels, 56
ephedrine, 155–156
ephedrine/caffeine/asprin stack, 156
Erpelding, Mark, *81*
every-other-day split, 52
exercise machines, 24–29
exercises, number and order of, 53

exercise style, 52
EZ bar curl, 129, 133, 135

F

fat, 71, 72, 73, 74, 75
faulty proportion, 178
Ferrigno, Lou, 49, 128
Fisher, Dave, *22*
Fitness, Johnny, 37, 96, 173
flat bench, 28, 59
flat bench lying dumbbell curl, 32
flat bench press, 99
flaxseed oil, 154–155
floor stretches, 39
Food for Sport, 71
forced reps, 32
forearms, exercises for the, 136–139, 188
Francois, Michael, *38, 68, 69*
free weights, 25–29, 53

G

Gaines, Charles, 175
Gironda, Vince, 49, 55, 64, 117, 124, 130, 164
Glass, Charles, *46*
glucose, 71
glutathione, 157
glycogen, 49, 70
goals, 10, 11, 12, 23, 45, 46
Gold's Gym, 33, 37, 49, 51, 60, 86, 165
good-morning exercise, 114
Grimek, John, *15,* 16, 63, 109, 123, 127, 157, 160–161, 181
Groulx, Claude, *78*
Guillaume, Paul-Jean, *94*
Gurley, Alq, *82*

H

Hack machine, 26
hack squat, 119, 120
hamstring stretch, 118
Haney, Lee, *11*
hanging leg raise, 107
Hardcore Bodybuilding, 49
hard gainer, 164–165
headaches, 188
heel-raise machine, 124
Hepburn, Doug, 63
high-fiber foods, 73
high-intensity workouts, 56
Hilligen, Roy, 127, 160
Hise, Jose, 161
home training, 159
Howlett, Stan, 6
hurdler's stretch, 118
hydroxy methylbutyrate (HMB), 149

I

incline barbell bench press, 100, 101
incline bench press, 101
incline dumbbell bench press, 99
incline dumbbell curl, 128, 130

incline flyes, 99
incline knee raise, 106
incline press, 88, 96
incline twisting sit-ups, 106
injuries, avoiding, 36–39, 182
inspiration, 46
insulin boosters, 150–156
inversion boot sit-ups, 107
Iron Man magazine, 11, 26, 35, 161, 169

J

Jablonicky, Pavol, *47, 80*
James, William, 46
Joe Weider's Flex magazine, 9
Jones, Arther, 26, 28, 173
joules, 189

K

Kern, Ericca, *141*
Key to Might and Muscle, 161
Kickinger, Roland, *46*
Koszewski, Zabo, 103, 104, 107, 160
Kruck, Debbie, *73*

L

Lannaghan, Darin, *146, 184*
Lasker, David, 176
lateral raise, 87, 88, 90, 92
lateral triceps, 187
lat machine, 25, 26
lat-machine pulldowns, 113
lat-pulldown apparatus, 26
lats, exercises for the, 109-115
lat-spread pose, 110
layoffs, 80
leg curls, 121
leg extensions, 119, 120
leg press, 26, 119, 121
Leidelmayer, Rory, 65
Levrone, Kevin, *34, 50, 87,* 128, *183*
lipids, 74
Long, Don, *76,* 128, *159*
low pulley rowing, 112
Lund, Chris, 83
lying triceps stretch, 135

M

MacFadden, Bernarr, 11
machines vs. weights, 24–29, 53
Maddron, Aaron, *20*
Makkawy, Mohamed, *103,* 128, 171, *172,* 177
Marek, Martin, 7
Matarazzo, Mike, *23, 85, 93*
Mayer, Dr. Jean, 72
McShane, Buster, 175
medial head, 87
melanocytes, 141
Mentzer, Mike *14,* 16, 31, 42, 56, 104, 133
mesomorph, 11, 15–19
metabolism, 66–69, 72
milk, 73–74

Miller, Joseph, 56
Milo course, 117
Mirkin, Dr. Gabe, 56
Mitos, Terry, 77
motivation, 8–12
Moyzan, Eddie, *81*
multi-poundage system, 161–162
Munzer, Andreas, *28, 84*
Muscle & Fitness magazine, 21, 26, 56, 176
muscle fiber, 31
MuscleMag International, 49, 60, 80, 104, 124
muscle size, adding, 158–166
"muscle sleep," 82–84
muscle tear, 37
music, 46, 175, 176

N

Nautilus equipment, 26, 27, 28, 29
Neary, Jack, 173
Neece, Steve, 173
New Age Magazine, 11
Nubret, Serge, *30,* 32, *83,* 84, *143*
Nubret pro rep method, 32
nutrition, 61, 64, 70–77, 181–182
nutritional overloading, 182

O

Oliva, Sergio, 109, 171
overcompensation, 21
overhead presses, 27
"overtonis," 55
overtraining, 42, 55, 80, 181, 188–189
Oxygen magazine, 60

P

Palumbo, David, *118*
parallel-bar dips, 27, 97, 128, 133
Park, Reg, 80, *95,* 109, 123, 127, *147,* 162, 176
partial reps, 32
Passariello, Ken, 171
Patterson, Bruce, *27, 85*
Pattyn, Tom, *54*
peak contraction, 32
Pearl, Bill, 104, 137
Pearson, Tony, 109
Peary Rader's routine, 161
Pek-Dek, 26, 27, 159
Physical Fitness of Champion Athletes, 63
pitfalls, 180–185
Platz, Tom, 42, *43,* 49, *117, 118,* 119
posing, 174–179
positive thinking, 12
posterior head, 87
Poulin, Jeff, *85*
Poulsen, Michael, *167*
power-building techniques, 52
"power walk," 124
preacher curl, 38, 129
pre-exhaust reps, 35, 49
press-behind-neck, 87, 88, 89
pressdowns on lat machine, 133

Price, Sue, *22*
Priest, Lee, *48,* 52, *67,* 68, *127,* 128, *137,*
   140, 180
progression, 51
progressive resistance exercise, 79
prone hyperextensions, 114
protein, 71, 72
pulling muscles, 33
pulse rate, 56
Pumping Iron, 175
pushing muscles, 33
pyramid training reps, 35

Q
questions and answers, 186–189
"quick gainers," 145

R
Rader, Peary, 161
Ray, Shawn, *34, 51,* 52, 103, *104,* 109, 177
recuperation, 54–61, 83
Reeves, Steve, 80, 103, 109, 124, 127, 162
Reg Park Journal, 162
Reg Park's routine, 162
relaxation, 59, 68
reps, 30–35, 52
rest, 79
rest-pause reps, 32
reverse curl, 138
reverse wrist curl, 138
Reynolds, Bill, 119
Robert, Lee, 103
Robinson, Eddie, 103, 171
Robinson, Robbie, 128, *135*
Roman chair sit-ups, 107
Ross, Clancy, 127
routines, 159–166

S
Sandow, Eugene, 63, 123, 178
Sarcev, Milos, *23, 53, 85,* 103, 105
Sarni, Mauro, *23, 177*
Schlierkamp, Gunther, 169
Schwarzenegger, Arnold, *10,* 11, 16, *41,*
   46, 52, *58,* 80, 90, 96, 107, 109, 123, 128,
   131, 137, 138, 144, *151,* 163, 164,
   169–170, 184
Schweitzer, Dr. Albert, 74
Scott, Larry, 90, 92, 124, 133, 137
Scott curl, 131
seated calf raise, 125
seated dumbbell press, 91
seated triceps dumbbell extensions, 134
sets, 189
Sheldon, Dr. William H., 15
Sherman, John, 175
"shock treatment," 80

shoulders, exercises for the, 86–92
shoulder tears, 38
shrugs, 114
Simmons, John, *96, 129*
single-arm dumbbell rowing, 113
single-arm dumbbell triceps stretch, 134
single-arm lateral raise, 91
single-arm triceps stretch, 39
Sipes, Chuck, 137
sissy squat, 121
sit-ups, 106, 107
skin, 141–142
sleep, 59, 83–84
Smith machine, 119
Smith, Dr. Nathan, 71
smoking and drinking, 182, 184
sodium, 74
somatotyping, 14–19
Sonbaty, Nasser El, *18,* 49, *51, 68,* 95, 128
specialization, 80
Spector, Dave, 38
speed of training, 52
Spinello, Joe, *65*
split routines, 163
Sports Medicine Book, 56
squats, 27, 38, 68, 117, 118, 119, 120, 121
squat stands, 159
standing calf raise, 123, 125
standing dumbbell curl, 132
standing groin stretch, 39
Steinborn, Henry, 161
Stern, Lee, 173
steroids, 57, 58, 59, 74, 101, 174
Steve Reeve's routine, 162
sticking point, 78–81
straight-set reps, 31
Strength magazine, 161
stretching exercises, 39
stretch marks, 189
strict reps, 32
Strydom, Gary, *45, 181*
sun and fresh air, 59–60
sunbathing, 142–143
suntan lamps, 143
super routine, 160
superset reps, 33
supine flying, 100
supplements, 22, 146–157

T
tanning pills, 144
tanning techinques, 140–145, 170
Tanny, Armand, 21, 127, 160
Taylor, Vince, *12,* 123, 124, 128, *173,*
   177, *179*
T-bar rowing, 38, 112
tendinitis, 38–39, 128

testosterone boosters, natural, 149–150
thigh-curl apparatus, 25
thigh-extension apparatus, 25, 26
thighs, exercises for the, 116–121
Tinerino, Dennis, 103
Titus, Craig, *37, 49, 141, 185*
training attitude, 53
training intensity, 50
training log, 20–23
traps, exercises for the, 109–115
tribulus terrestris, 150
triceps, exercises for the, 27, 128, 133–135
Turner, George, 26

U
undergrip close-handed chin, 132
undernourishment, 181
upright rowing, 27, 88, 89

V
vanadyl sulphate, 152
Varga, Tom, *122*
veins, 188
Vigor magazine, 161
visualization, 11–12, 45–46, 81
vitamins, 64–65, 156–157

W
Walker, Roger, *10*
warming up, 37
Weider, Ben, *42, 58*
Weider, Joe, 9, 25, *42,* 169, 170
weight, amount of, 53
Wheeler, Flex, *13,* 45, *55,* 56, *57, 61,*
   108, 123, 124, 128, *166,* 169, *171, 174,*
   177, *178*
Wheeler, Ken, 173
whey protein, 154
wide-grip chin, 26, 110, 112, 113, 115
wide-grip lat-machine pulldown, 110, 112
wide-grip upright row, 88
Wilson, Scott, *6*
workout duration, 80
World's Gym, 49, 51, 165
wrist curl, 138, 139
wrists, building, 187

Y
Yates, Dorian, 16, *17,* 18, *21, 42,* 56, *57,*
   111, 109, 124, *153,* 166
Yorton, Chet, 19

Z
Zane, Frank, *9, 16,* 21, 64, 104, *169,*
   170, 177
Zarco, Hank, 60
Zulak, Greg, 173